Measurement, Probability, Graphs, and Geography

David Eastwood

PARTRIDGE

Copyright © 2019 by Brain Based Education.
Math Without Calculators
Algebra Two
Medina, Ohio USA

ISBN: Softcover   978-1-5437-0555-3

All rights reserved. No part of this book may be used or reproduced by any means, graphic, electronic, or mechanical, including photocopying, recording, taping or by any information storage retrieval system without the written permission of the author except in the case of brief quotations embodied in critical articles and reviews.

Summary: "This book Is for students who want to learn math, especially those who want to learn it without calculators. It is filled with suggestions and problems to review it."

Because of the dynamic nature of the Internet, any web addresses or links contained in this book may have changed since publication and may no longer be valid. The views expressed in this work are solely those of the author and do not necessarily reflect the views of the publisher, and the publisher hereby disclaims any responsibility for them.

Print information available on the last page.

To order additional copies of this book, contact
Partridge India
000 800 10062 62
orders.india@partridgepublishing.com

www.partridgepublishing.com/india

1. Measurement 2. Graphs 3. Probability and 4. Geometry

1. Measurement

1. Measurements Less than 1 Km.....1

1-1	Inch, Feet, and Yard	1
1-2	Change feet to yards	3
1-3	Meters, Centimeters	5
1-4	Millimeters, Centimeters	7
1-5	Fractions of an Inch	9
1-6	Review Problems	11

2. Meters, Centimeters, and Km......13

2-1	Miles and 10ths of a Mile	13
2-2	Kilometers	15
2-3	Story Problems	17
2-4	Review Problems	19

3. Measuring Liquids........................21

3-1	Cup, Quart, and Gallon	21
3-2	Liter, Centiliter, and Deciliter	23
3-3	Milliliter and Kiloliter	25
3-4	Teaspoon	27
3-5	Review Problems	29

4. Measuring Weight/Temperature...31

4-1	Pound and Ounce	31
4-2	Tons and 10ths of a Ton	33
4-3	Grams and Kilograms	35
4-4	Temperature	37
4-5	Review Problems	39

5. Range, Average, Mode, Median...41

5-1	Range and Close Range	41
5-2	Average	42
5-3	Median	43
5-4	Mode	44
5-5	Review Problems	45

2. Graphs

6. Basic Graphs............................51

6-1	Bar and Picture Graphs	51
6-2	Time on a Graph	53
6-3	Frequency/ Stem Graphs	55
6-4	Surveys and Percents	57
6-5	Review Problems	59

7. Graphs Part 2..........................61

7-1	Circle Graphs	61
7-2	Venn Diagrams	63
7-3	Number Grid Graphs	65
7-4	Making Graphs	67
7-5	Review Problems	69

3. Probability

8. Probability of 1 Event..............71

8-1	Probability Words	71
8-2	Fractions of a Event	73
8-3	Percent of an Event	75
8-4	Review Problems	77

9. Probablity of Multiple Events...79

9-1	Use an Exponent	79
9-2	Dependent Events	81
9-3	Permutations	83
9-4	Combinations	85
9-5	Review Problems	87

4. Geometry

10. Basic Angles...........................89

10-1	Lines and Degrees	89
10-2	30 and 45 Degree Angles	91
10-3	Directions of a Compass	93
10-4	Complement/Supplement	95
10-5	Review Problems	97

11. Shapes and How They Move 99

- 11-1 Triangles/ Quadrilaterals   99
- 11-2 More than 4 Sides   101
- 11-3 Symmetry   103
- 11-4 Congruency and Translation   105
- 11-5 Review Problems   107

12. Perimeter and Area 109

- 12-1 Perimeter/ Area   109
- 12-2 American Square Units   111
- 12-3 Metric Square Units   113
- 12-4 Area of a Triangle/ Trapezoid   115
- 12-5 Review Problems   117

13. Solids and Volume 119

- 13-1 Volume of a Box   119
- 13-2 Surface Area of a Box   121
- 13-3 Surface Area Triangle   123
- 13-4 Volume Triangle/ Pyr   125
- 13-5 Review Problems   127

14. Measuring with Circles 129

- 14-1 Perimeter of a Circle   129
- 14-2 Area of a Circle/ Cylinder   131
- 14-3 Surface Area Can/ Cone   133
- 14-4 Volume of a Can/ Cone   135
- 14-5 Sphere   137
- 14-6 Review Problems   139

Ch 4 Ls 1: 2 Steps to Carry Addition   33

_____ Front ____ / 8   Back ____ / 27   Rev ____ / 20   T / 53 _____
         Name   Checker

#1 1. When do you carry in math? _____
```
         26
```
2. What does 6 + 6 carry? **+ 6** _____   **Checker makes**
                                                               sure it's done.
3. What's the 2nd step to carry? _____

4. Why don't you have to carry twice with 100s? _____

#2 1. You know 3 + 9 is 12.   13
    What's the 1st step to carry?   + 9   **Student makes**
                                       ———   **sure these are**
                                        2   **filled out in class.**

                                      → ¹1 3
              Carry a ____.   + 9
                                       ———
                                        2

                                      tens ones
                                        15   **Checker reviews**
    2. What's the teen fact?   + 9   **the front page.**
                                       ———

Why we're different!!!

                            4 + 8 = ____

                             ___ ___

                              26
    4. What's the teen fact? + 7
                             ———

                            6 + 7 = ____

                             ___ ___

Student is honest about whether they used a calculator or not. → Calculator? yes no

Review 1. When do you carry in math? _____ Calculator? yes no
2. Name 2 steps to carry. _____
3. Why don't you have to carry twice with 100s? _____

Student is quizzed on these Qs.
(Teacher option)

It works!!!

PLUS These explore these pages...

1. What's Happening in Algebra The student decides what's happening in algebra.

2. Simplified Algebra means it's easier to make equations, not just solve equations someone else has made.

3. Plenty of story problems!!!

It's written as Math Without Calculators, but you need to make it happen.

**You can do this!
Mr David Eastwood**

Rules Sheet

As you go through the lessons you will see some rules that you are unfamiliar with, so I developed this paper to help you understand them.

1. Ma Rule: The intials are from the rule itself, "Multiply same bases, Add the exponents." You probably know the Product of Powers Property.
   Example: $2^2 \times 2^4 = 2^6$

2. Me Rule: These initials are also from the rule, "Multiply the numbers, Exponents stay the same." (E shows that exponents stay the same.)
   Example: $3^4 \times 5^4 = 15^4$

3. DS Rule: Again, the initials show the rule. "Divide same bases, Subtract the Exponents."
   Example: 4^5 divided by $4^2 = 4^3$

4. Simpliify: No need for initials here, just simplify.
   Example: 20^3 divided by $4^3 = 5^3$

5. Subtraction: 2. Count Up Subtraction: Example: 23 - 9   Count Up 9 to 10 and subtract, it's 13. Add the 1 from the bottom. It's 14.

These are the name changes up until now. Any others?

Ch 1 Ls 1 Inches, Feet, and Yards . 1

_____ #1 #2 ____/12 #3 #4 ____/ 16 R ___/ 20 T ____/ 48 _____
           Name   Checker

#1 1. Find these facts. 1 foot = ___ inches   1 yards = ___ feet   1 yard = ___ inches

2. What math fact changes 4 yards into feet? _____

3. What fact changes 6 yards to feet?   **6 yards = ? feet**

4. How do you change 6 yards?   **1 yard is ___ feet**

      Multiply. 6 x ___ = ___   6 yards = ___ feet

5. Change yards to feet.   **4 yards = ? feet**   **4 feet = ? inches**

                    x ___ = ___ feet   4 x ___ = ___ feet

#2 1. What's the 1st step to change 4 yards 2 feet into feet? _____

2. What's the 2nd step to make it all feet? _____

3. Name 2 steps to change yards and feet into just feet. _____

4. What multiplies and what adds?   **5 yards 1 ft = ? feet**

      How many feet is that?   Multiply ___ x ___ and add ___ feet

                                      5 yards 1 ft = ___ feet

5. What multiplies and what adds?   **2 feet 1 inch = ? inches**

      How many feet is that?   Multiply ___ x ___ and add ___ inch

                                      5 ft 1 in = ___ inches

6. All 1 step. How many feet is it?   **7 yards 2 ft = ? feet**

                                      7 yards 2 ft = ___ feet

#3 Solve these basic facts.   Calculator? yes no

1. 3 yds = ___ feet   5 yds 1 ft = ___ feet
2. 4 feet = ___ inches   6 ft 7 inch = ___ inches
3. 5 yds = ___ feet   9 yds 2 ft = ___ feet
4. 7 feet = ___ inches   4 ft 8 inch = ___ inches

#4 Is it correct? Yes or correct.   Calculator? yes no

1. 3 yds 1 ft = 11 feet   yes or _____ ft   4 yd 1 ft = 13 ft   yes or _____ ft
2. 7 yds 2 ft = 23 feet   yes or _____ ft   8 yds 2 ft = 25 ft   yes or _____ ft
3. 4 ft 9 inch = 55 in   yes or _____ in   6 ft 7 in = 78 in   yes or _____ in
4. 5 ft 4 inch = 64 in   yes or _____ in   3 ft 10 in = 46 in   yes or _____ in

Review 1. Find these facts. 1 foot = ___ inches   1 yards = ___ feet   1 yard = ____ inches   Calculator? yes no
2. What math fact changes 4 yards into feet? _____
3. What's the 1st step to change 4 yards 2 feet into feet? _____
4. What's the 2nd step to make it all feet? _____
5. Name 2 steps to change yards and feet into just feet. _____

6.   3 yards = _____ feet   5 feet = _____ inches
7.   4 feet = _____ inches   9 yards = _____ feet
8.   5 yards 1 ft = _____ feet   2 yards 2 ft = _____ feet
9.   4 ft 7 inch = _____ inches   6 ft 8 inch = _____ inches
10. 6 yards 2 ft = _____ feet   7 yards 1 ft = _____ feet
11. 3 ft 10 inch = _____ inches   5 ft 3 inch = _____ inches
12. 4 yards 2 ft = _____ feet   6 yards 2 ft = _____ feet
13. 8 yd 3 ft = _____ feet   4 yd 5 inches = _____ inches

Each lesson has a quiz.

Ch 1 Ls 2 Change Units with Division.   3

_____ #1 #2 ____/ 12 #3 #4 ____/ 18 R ___/ 19 T ____/ 49 _____
                Name   Checker

#1 1. What basic fact changes 12 feet into yards? _____
    2. How does a basic fact show to divide? _____
    3. What fact changes feet to yards?   **15 feet = ? yards**

      How can you tell it's divide? Solve it.   **1 yard is ___ feet**

      Divide because _____ . 15 ÷ ___ = ___   **15 feet = ___ yds**

    4. How many yards are in 12 feet?   **12 feet = ? yards**

                                    ___ ÷ ___ = ___ yards

#2 1. What's the 1st step to change 13 feet to yards and ft? _____
    2. What's the 2nd step to change 13 feet? _____
    3. What are the 2 steps together? _____
    _____

    4. What 2 steps does it use?   **17 feet = ? yards ? feet**

      How many yards and feet is that?   **Divide 17 by ___ and subtract ___ ft**

                                17 feet = ___ yards ___ feet

    5. What 2 steps does it use?   **25 feet = ? yards ? feet**

      How many yards and feet is that?   **Divide 24 by ___ and subtract ___ ft**

                                25 feet = ___ yards ___ feet

    6. What 2 steps does it use?   **11 feet = ? yards ? feet**

                                11 feet = ___ yards ___ feet

#3 Solve these basic facts.   Calculator?
                                                                           yes   no

1. 26 inches = ____ feet ____ in   27 inches = ____ ft ____ in
2. 64 inches = ____ feet ____ in   68 inches = ____ ft ____ in
3. 14 feet = ____ yds ____ ft   19 feet = ____ yds ____ ft
4. 53 inches = ____ feet ____ in   55 inches = ____ ft ____ in
5. 40 feet = ____ yds ____ ft   26 feet = ____ yds ____ ft

#4 Is it correct? Yes or correct.   Calculator?
                                                                           yes   no

1. 19 in = 2 ft 3 in   yes or _____   86 in = 7 ft 2 in   yes or _____
2. 14 ft = 3 yds 1 ft   yes or _____   17 ft = 5 yds 1 ft yes or _____
3. 28 in = 2 ft 4 in   yes or _____   76 in = 5 ft 2 in   yes or _____
4. 29 ft = 9 yds 2 ft   yes or _____   20 ft = 6 yds 2 ft yes or _____

Review 1. What basic fact changes 12 feet into yards? _____   Calculator?
                                                                                 yes   no
2. How does a basic fact show to divide? _____
3. What's the 1st step to change 13 feet to yards and ft? _____
4. What's the 2nd step to change 13 feet? _____
5. What are the 2 steps together? _____
   _____

6.   36 inches = ____ feet   63 inches = ____ feet ____ in
7.   15 feet = ____ yards   17 feet = ____ yards ____ ft
8.   30 inches = ____ ft ____ in   92 inches = ____ ft ____ in
9.   26 feet = ____ yds ____ ft   40 feet = ____ yds ____ ft
10. 43 inches = ____ ft ____ in   78 inches = ____ ft ____ in

Ch 1 Ls 3 Use fractions of an inch. 5

_____ #1 #2 ____ / 11 #3 ____ / 5 R ____ / 8 T ____ / 24 _____
           Name   Checker

#1 1. What units divide an inch into 2 and 4 parts? _____
2. What units divide an inch into 8 and 16 parts? _____
3. How do you put whole inches with fractions? _____
4. Do smaller units make more or less accurate answers? _____
5. What fraction of an inch is it?

$\dfrac{}{8}$ inch

6. What fraction is it?

$\dfrac{}{8}$ inch

7. What fraction is it?

$\dfrac{}{16}$ inch

#2 1. How do fractions work with whole inches? _____
2. How long is this bar?

____ $\dfrac{}{2}$ inches

3. How long is this bar?

____ $\dfrac{}{4}$ inches

4. What mixed number is it?

____ $\dfrac{}{4}$ inches

6.

#3 Solve story problems with fractions of inches.   Calculator?
                                                                       yes   no

1. Lumber is 1 1/2 inches thick. JT's truck can
   carry 18 inches. How many sets of lumber   _____
   can he carry in a load?

2. Mitul has a 50 piece puzzle, but there's
   only 45 pieces. What fraction is he missing?   _____

3. It's 20 and 1/2 ft of sidewalk from the garage
   to the house. JT has enough concrete for   _____
   16 3/4 ft. How much does he need to finish it?

4. Mrs J has 6 dozen eggs. She has enough
   ommeletes 4 1/3 eggs. What's left?   _____

5. Ojas won the long jump with 15 ft 4 3/8 in. He
   needs 1 ft 2 2/8 in to get the record. How   _____
   long does he need to get it?

Review 1. What units divide an inch into 2 and 4 parts? _____   Calculator?
                                                                                      yes   no
2. After 4ths, what is the next way it's divided into? _____

3. What fraction unit is after 8ths? _____

4. Do smaller units make more or less accurate answers? _____

1. How long is a centimeter compared to an inch? _____

2. Two inches equals how many centimeters? _____

How long are these measurements?

5. [ruler 1-5, arrow at 4]   ___ / 4 inches

6. [ruler 1-6, arrow at 5-6]   ___ / 4 inches

7. [ruler 0-1, arrow near middle] ___ / 16 inches

8. [ruler 0-1, arrow]   ___ / 16 inches

Ch 1 Ls 4 Find units of measurement. 7

_____ #1 #2 ____/ 11 #3 #4 ____/ 16 R ___/ 19 T___/ 46 _____
Name   Checker

#1 1. Which is longer, a yard or meter, and by how much? _____
2. Which metric unit is smaller than an inch? _____
3. If meters are like dollars, what are centimeters? _____
4. What unit is alot like dimes? _____
5. How do you change to centimeters?   **4 meters = ? cm**

Multiply meters x ____   4 x 100 = ____   4 meters = ____ cm
Find each answer.

6. How many centimeters is 1.2 meters?   **1.2 meters = ? cm**

4 x 100 = ____   1.2 meters = ____ cm

7. How many meters is 450 centimeters?   **450 cm = ? meters**

450 cm is ____ meters

#2 1. How long is a centimeter compared to an inch? _____
2. Two inches equals how many centimeters? _____
3. How do you count cm in these inches?   **6 inches**

Count the 2s. How many cm?   ____ in = 5 cm

6 inches is ____ cm

4. How many cm is 14 inches?   **14 inches**

14 inches is ____ cm

5. How many cm is 20 inches?   **20 inches**

20 inches is ____ cm

8.

#3 Change with centimeters and decimeters.   Calculator?
                                                     yes   no

1. 4 meters = _____ cm   5.8 meters = _____ cm

2. 6 meters = _____ cm   3.7 meters = _____ cm

3. 320 cm = ___ m ___ cm   170 cm = ___ m ___ cm

4. 280 cm = ___ m ___ cm   715 cm = ___ m ___ cm

5. 482 cm = ___ m ___ cm   678 cm = ___ m ___ cm

#4 Change with centimeters and inches.   Calculator?
                                                     yes   no

1. 8 inches is _____ cm   6 inches is _____ cm

2. 12 inches is _____ cm   16 inches is _____ cm

3. 24 inches is _____ cm   30 inches is _____ cm

Review 1. Which is longer, a yard or a meter, and by how much? _____   Calculator?
                                                                                yes   no
2. Which metric unit is smaller than an inch? _____

3. If meters are like dollars, what are centimeters? _____

4. What unit is alot like dimes? _____

5. How long is a centimeter compared to an inch? _____

6. Two inches equals how many centimeters? _____

Change these.

7.   1.4 meters = _____ cm   2.6 meters = _____ cm

8.   2.1 meters = _____ cm   3.5 meters = _____ cm

9.   420 cm = ___ m ___ cm   530 cm = ___ m ___ cm

10. 320 cm = ___ m ___ cm   130 cm = ___ m ___ cm

11. 4 meters = _____ dm   7 meters = _____ dm

12. 15 dm = _____ m   26 dm = _____ m

13. 42 dm = _____ m   76 dm = _____ m

14. 8 meters = _____ cm   7 dm = _____ cm

Ch 1 Ls 5 Go backwards to change units. 9

_____ #1 #2 ____/ 10 #3 #4 ____/ 18 R ____/ 15 T ____/ 58 _____
          Name   Checker

#1 1. What metric unit is smaller than a centimeter? _____

2. How do you change 3 meters to millimeters? _____

3. 500 millimeters is how many meters? _____

4. How do you change millimeters to meters? _____

5. How do you change to millimeters?   **5 meters = ? mm**

        Multiply meters x ____   5 x 1000 = ____   5 meters = ____ mm
        Find each answer.

6. How many centimeters is 1.3 meters?   **1.3 meters = ? mm**

        1,000 + 300 = ____   1.3 meters = ____ mm

7. How many meters is 600 millimeters?   **600 mm = ? meters**

        600 mm is ____ meters

#2 1. A centimeter is how many millimeters? _____

2. 4 centimeters is how many millimeters?   **4 cm = ? mm**

        Each cm is ___ mm.   4 cm = ____ mm

3. 53 centimeters is how many millimeters?   **53 cm = ? mm**

        53 cm = ____ mm

4. 45 millimeters is how many centimeters?   **45 mm = ? cm**

        45 mm = ____ cm

5. 2 millimeters is how many centimeters?   **2 mm = ? cm**

        2 mm = ____ cm

10.

#3 Change with meters and millimeters.

Calculator? yes no

1. 6 meters = _____ mm   4.6 meters = _____ mm
2. 5 meters = _____ cm   5.2 meters = _____ cm
3. 120 cm = _____ m   230 cm = _____ m
4. 1030 mm = _____ m   3700 mm = _____ m
5. 4200 mm = _____ m   2300 mm = _____ m

#4 Is it correct? Yes or correct.

Calculator? yes no

1. 4 cm = 400 mm   yes or _____   7 cm = 70 mm   yes or _____
2. 1.2 cm = 12 mm   yes or _____   1.7 cm = 17 mm   yes or _____
3. 2.7 cm = 270 mm yes or _____   8.9 cm = 89 mm   yes or _____
4. 320 mm = 32 cm   yes or _____   130 mm = 1300 cm yes or _____

Review 1. What metric unit is smaller than a centimeter? _____

Calculator? yes no

2. How do you change 3 meters to millimeters? _____
3. 500 millimeters is how many meters? _____
4. How do you change millimeters to meters? _____
5. A centimeter is how many millimeters? _____

6. 6 m = _____ mm   7 m = _____ mm
7. 1.2 m = _____ cm   1.6 m = _____ cm
8. 320 mm = _____ cm   130 mm = _____ cm
9. 4 cm = _____ mm   5 cm = _____ mm
10. 5.3 cm = _____ mm   6.2 cm = _____ mm
11. 630 mm = _____ cm   750 mm = _____ cm
12. 2.3 cm = _____ mm   5.7 cm = _____ mm

Problems Review   11

_____ #1 #2 #3 ____/ 41 #4 #5 ____/ 12 Total ____/ 53
                Name

#1   1. Meter _____
     2. Centimeter _____
     3. Decimeter _____
     4. Millimeter _____

   #2 1.   1.7 meters = _____ cm   3.4 meters = _____ cm   Calculator?
       2.   390 cm = ___ m ___ cm   570 cm = ___ m ___ cm   yes   no
       3.   650 cm = ___ m ___ cm   860 cm = ___ m ___ cm
       4.   4 meters = _____ dm   6.4 meters = _____ dm
       5.   21 dm = _____ m   37 dm = _____ m
       6.   7 m = _____ mm   8.2 m = _____ mm
       7.   2.3 m = _____ cm   4.5 m = _____ cm
       8.   580 mm = _____ cm   670 mm = _____ cm
       9.   5 cm = _____ mm   44 cm = _____ mm
       10. 40.6 cm = _____ mm   48.6 cm = _____ mm

#3 1. Inch _____
    2. Foot _____
    3. Yard _____

#4 1.   64 inches = ___ feet ___ inches   45 inches = ___ feet ___ in   Calculator?
    2.   53 inches = ___ ft ___ in   64 inches = ___ ft ___ in   yes   no
    3.   71 inches = ___ ft ___ in   86 inches = ___ ft ___ in
    4.   19 feet = ___ yds ___ ft   25 feet = ___ yds ___ ft
    5.   22 feet = ___ yds ___ ft   35 feet = ___ yds ___ ft
    6.   13 feet = ___ yds ___ ft   41 feet = ___ yds ___ ft
    7.   3 yds ft 2 = ___ feet   4 yds 1 ft = ___ feet
    8.   5 yds ft 1 = ___ feet   7 yds 2 ft = ___ feet

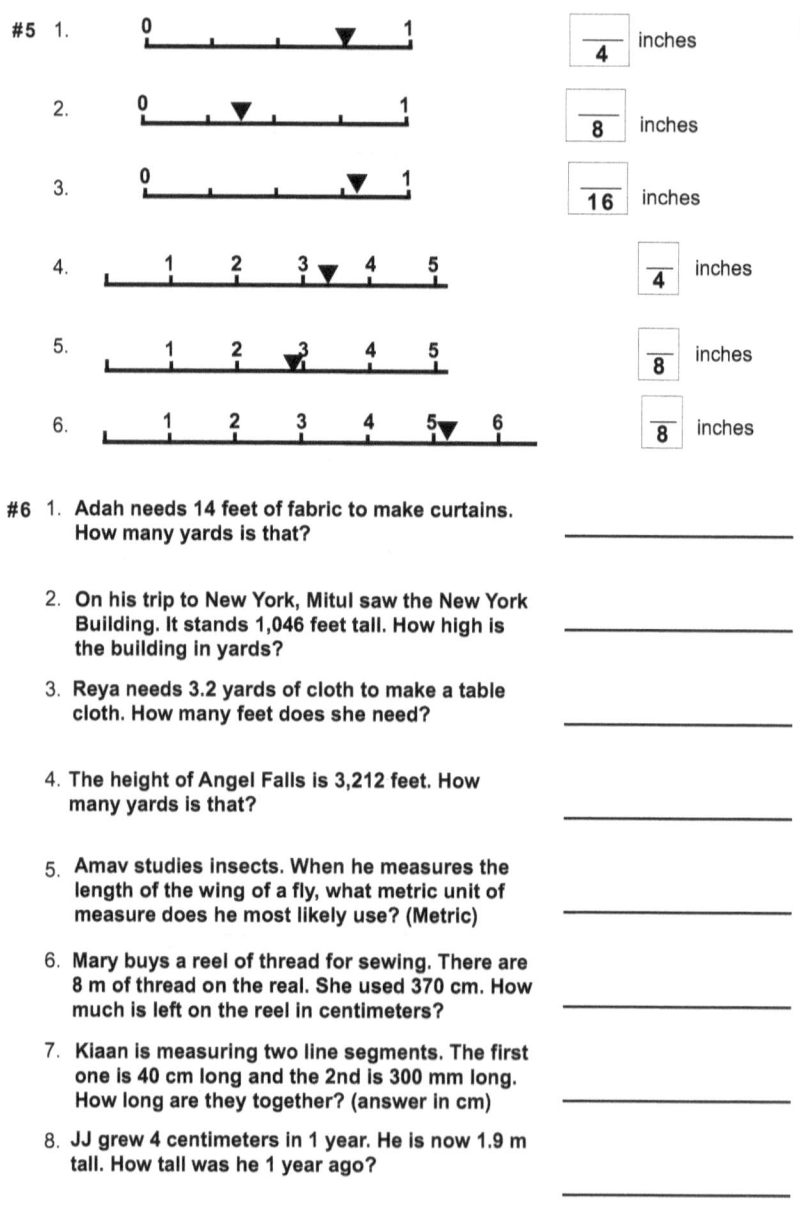

#5
1. ___/4 inches   Calculator? yes no
2. ___/8 inches
3. ___/16 inches
4. ___/4 inches
5. ___/8 inches
6. ___/8 inches

#6
1. Adah needs 14 feet of fabric to make curtains. How many yards is that?

2. On his trip to New York, Mitul saw the New York Building. It stands 1,046 feet tall. How high is the building in yards?

3. Reya needs 3.2 yards of cloth to make a table cloth. How many feet does she need?

4. The height of Angel Falls is 3,212 feet. How many yards is that?

5. Amav studies insects. When he measures the length of the wing of a fly, what metric unit of measure does he most likely use? (Metric)

6. Mary buys a reel of thread for sewing. There are 8 m of thread on the reel. She used 370 cm. How much is left on the reel in centimeters?

7. Kiaan is measuring two line segments. The first one is 40 cm long and the 2nd is 300 mm long. How long are they together? (answer in cm)

8. JJ grew 4 centimeters in 1 year. He is now 1.9 m tall. How tall was he 1 year ago?

Ch 2 Ls 1: Use miles and 10ths of a mile. 13

_____ #1 #2 ____/ 11 #3 #4 ____/ 16 R ___/ 17 T ____/ 47 _____
      Name   Checker

#1   1. How many feet are in a mile? _____

        2. How do you round to thousands? _____

        3. How many feet are in a 10th of a mile? _____

        4. How do you estimate 20,000 feet in miles? _____

        5. Use 5000 ft per milee. Estimate 1500 ft in mi.   **15,000 ft**

                                                                     _____ miles

        6. Estimate 45,000 ft in miles.   **45,000 ft**

                                                                   _____ miles

#2   1. Use 5000 ft for a mile. How many feet are in a half mile? _____

        2. How many feet are in a 10th mile? _____

        3. Estimate 10ths of a mile. What does it multiply?   **0.3 miles**

            Multiply each 10th by _____. How many feet?   3 x _____ = ?

                                                              _____ ft

        4. Estimate it. What does it multiply?   **0.8 miles**

            Multiply each 10th by _____. How many feet?   8 x _____ = ?

                                                              _____ ft

        5. How do you estimate 10ths of a mile?   **4000 ft**

            Count the 500s. How many 10ths is it?   **4000 has ____ 500s**

                                                    4000 ft is about _____ mile.

#3 Estimate 10ths of a mile using 500 feet.   Calculator? yes no

1. 0.3 mi = _____ ft   0.7 mi = _____ ft
2. 0.4 mi = _____ ft   0.2 mi = _____ ft
3. 1.4 mi = _____ ft   2.6 mi = _____ ft
4. 3.1 mi = _____ ft   4.5 mi = _____ ft

#4 Use 5000 feet for a mile. Correct or change.   Calculator? yes no

1. 15,000 ft = 3 mi   yes or _____   7,500 ft = 2 mi   yes or _____
2. 20,000 ft = 5 mi   yes or _____   35,000 ft = 7 mi yes or _____
3. 45,000 ft = 9 mi   yes or _____   12,500 ft = 2 mi yes or _____
4. 30,000 ft = 5 mi   yes or _____   41,000 ft = 8 mi yes or _____

Review 1. How many feet are in a mile? _____   Calculator? yes no

2. How do you round a mile to thousands? _____
3. How do you estimate 20,000 feet in miles? _____
4. Use 5000 ft for a mile. How many feet are in a half mile? _____
5. How many feet are in a 10th mile? _____

6. 16,000 ft = _____ mi   7,500 ft = _____ mi
7. 21,500 ft = _____ mi   38,000 ft = _____ mi
8. 17,500 ft = _____ mi   9,500 ft = _____ mi
9. 0.6 mi = _____ ft   0.7 mi = _____ ft
10. 1.8 mi = _____ ft   2.7 mi = _____ ft
11. 3.4 mi = _____ ft   1.9 mi = _____ ft

Ch 2 Ls 2 Use miles and kilometers 15

_____ #1 #2 ____/ 10 #3 #4 ____/ 18 R ___/ 19 T ____/ 47 _____
         Name   Checker

#1 1. How many meters are in a kilometer? _____

2. How do you change kilometers to meters?   **4 km = ? meters**

   Multiply x _____ .   4 x 1000 = _____   4 km = _____ meters

3. 4.7 km is how many meters?   **4.7 km = ? meters**

   4.7 km = _____ meters

4. 2,600 meters is how many km?   **2,600 meters = ? km**

   2,600 meters = _____ km

#2 1. What part of a mile is 1 kilometer? _____

2. 10 kilometers is about how many miles? _____

3. How do you change kilometers to miles?   **20 km = ? miles**

   Count 10s. Each 10k is _____ miles.   What do you multiply by?

   2 x 6 = _____ miles

4. 60 kilometers is how many miles?   **60 km = ? miles**

   _____ x 6 = _____ miles

5. 110 kilometers is how many miles?   **110 km = ? miles**

   _____ + _____ = _____ miles

6. 130 kilometers is how many miles?   **130 km = ? miles**

   _____ + _____ = _____ miles

#3 Change kilometers with meters.   Calculator? yes no

1. 5 km = _____ meters   4.5 k = _____ meters
2. 20 km = _____ meters   50 k = _____ meters
3. 80 km = _____ meters   90 k = _____ meters
4. 100 km = _____ meters   150 k = _____ meters

#4 Change kilometers with miles.   Calculator? yes no

1. 10 km = _____ miles   20 km = _____ miles
2. 50 km = _____ miles   70 km = _____ miles
3. 100 km = _____ miles   150 km = _____ miles
4. 200 km = _____ miles   250 km = _____ miles
5. 300 km = _____ miles   500 km = _____ miles

Review 1. How many meters are in a kilometer? _____   Calculator? yes no

2. What part of a mile makes 1 kilometer? _____

3. 10 kilometers is about how many miles? _____

4. 2 km = _____ meters   2.5 k = _____ meters
5. 6 km = _____ meters   6.5 k = _____ meters
6. 8 km = _____ meters   8.5 k = _____ meters
7. 10 km = _____ meters   30 km = _____ meters
8. 40 km = _____ miles   90 km = _____ miles
9. 100 km = _____ miles   120 km = _____ miles
10. 220 km = _____ miles   350 km = _____ miles
11. 400 km = _____ miles   600 km = _____ miles

Ch 2 Ls 3   Measurement story problems

_____ #1 #2 ____/ 8   #3 #4 ____/ 9   R ____/ 3 T ____/ 20 _____
Name   Checker

#1 1. What is important in measurement story problems? _____
2. How do you know which unit to use? _____
3. What 3 steps do all story problems follow? _____

#2 1. Find the key words and answer.

Reya's coach had her run 5 kilometers in the morning and 10 km at night. How many miles in all?

_____ _____
Key words   Answer

2. Key words? Answer it.

The carpet roll is 13 yards. They need 40 feet of carpet. Is it long enough?

_____ _____
Key words   Answer

3. Key words? Answer it.

A unit is 24 mm long. It needs to replace one that is 2.4 cm long. Does it fit?

_____ _____
Key words   Answer

4. Key words? Answer it.

The airport has a main runway that is 0.7 miles long. Jim's Jet takes 4500 ft to land. Is it enough?

_____ _____
Key words   Answer

5. Key words? Answer it.

Ojas needs a pencil that is 4 inches long. Keya had one that is 10 cm long. Is it long enough?

_____ _____
Key words   Answer

6. Key words? Answer it.

Mitul has 120 km for the trip. He's driven 45 miles. How far does he need to go?

_____ _____
Key words   Answer

#3 Solve measurement story problems.   Calculator?
                                                          yes   no

1. You've decided to plant tomatoes in your garden. Each tomato plant takes 6 inches and you have 8 yds to fill. How many tomatoes do you need? _____

2. You are making a poster and need 3 meters of string to go around it. You have 280 cm. Do you have enough to finish the poster? _____

3. The shortest person in class is 1.2 meters tall. If he jumps up he can reach 210 cm. How far can he reach? _____

4. A CD case is a quarter inch thick. How many inches long is 28 CDs? _____

5. Divit can jump and reach to 6 ft. 8 in. The hoop is 10 feet high. How close is he to reaching it? _____

6. TJ has 180 kiometers to drive. His car gets 20 kilometers per liter. How many liters does he need? _____

7. Amav's Jet takes 0.6 mi to land. The airport has 4000 ft. Is it enough for the jet to land? _____

8. Keya's pencil is 5 inches long. Sana's pencil is 13 cm long. Which pencil is longer? _____

9. A basketball court is 25 yards. How many rolls of 10 feet tape is that? _____

Review 1. What is important in measurement story problems? _____

2. How do you know which unit to use? _____

3. What 3 steps do all story problems follow? _____

_____

Review Problems 19

_____ #1 #2 #3 ____/37 #4 #5 ____/ 13 T ____/30
Name

1. Mile _____

2. Kilometer _____

#3 How long is each measurement?   Calculator?
                                                      yes   no

1.   7,000 ft = _____ miles   16,500 ft = _____ mi

2.   17,500 ft = _____ mi   26,000 ft = _____ mi

3.   22,500 ft = _____ mi   10,500 ft = _____ mi

4.   18,000 ft = _____ mi   9,500 ft = _____ mi

5.   0.7 mi = _____ ft   1.3 mi = _____ ft

6.   2.5 mi = _____ ft   3.4 mi = _____ ft

7.   4.2 mi = _____ ft   5.5 mi = _____ ft

8.   3.5 mi = _____ ft   6.2 mi = _____ ft

#4 Change kilometers with miles.

1.   3.5 km = _____ meters   4.7 k = _____ meters

2.   6.8 km = _____ meters   7.5 k = _____ meters

3.   8.2 km = _____ meters   9.3 k = _____ meters

4.   20 km = _____ meters   50 km = _____ meters

5.   60 km = _____ miles   80 km = _____ miles

6.   110 km = _____ miles   140 km = _____ miles

7.   200 km = _____ miles   380 km = _____ miles

8.   500 km = _____ miles   700 km = _____ miles

#5

Solve these story problems.

1. Ojas rode 2 kilometers on his bike, while Ira rode 2500 meters. Who rode the farthest and how much farther? (answer is in meters) _____   Calculator? yes   no

2. Mitul wants to send a box of oranges. The box cannot exceed 20 kg. Each orange has a mass of 200 g, what is the maximum number she can send? _____

3. Adah lives 3/4 of a mile from school while JJ lives 0.6 miles away. How much farther does she walk? _____

4. Hiran is moving from Delhi, India, to Cleveland, Ohio. Delhi's annual rainfall is 2.8 feet, while Cleveland's is 40 inches. Which city gets more rain? _____

5. Kiaan measured a line for his history project. It is 4.5 meters long. How many centimeters is it? _____

6. Anvi's box is 22 inches long and 9 inches wide. How many more inches is it longer than it is wide? _____

7. Eva's old house is 2.3 miles from her new house. How many feet is that? (rounded) _____

8. Mr K has a 1.6 yard long piece of wood that he wants to cut it into 4 equal lengths. How long should each piece be in inches? _____

9. It's 3.7 miles to town from Mitul's house. How many feet is that? (Round it.) _____

10. Hansh heard that their jet was cruising at 35,000 feet. About how high is that in miles? _____

11. Mr T's jet needs 0.7 of a mile to land. He radioed that the airstrip was 4000 feet. Is it safe to land? _____

12. Pari's box is 45 cm long and 18 cm wide. How many more cm is it longer than wide? _____

Ch 3 Ls 1 Cups, Quarts, and Gallons 21

_____ #1 #2 ____ / 10 #3 #4 ____ / 16 R ___ / 14 T____ / 40 _____
                Name   Checker

#1 1. How many cups are in a can of 12 ounce pop? _____

2. A quart is how many cups? _____   3. A gallon is how many quarts? _____

4. What number compares cups, quarts, and gallons? _____

5. How many cups are in 1 quart?   **2 quarts = ? cups**

Do you multiply or divide?   **1 quart = ____ cups**   How many in all?

Circle one   Multiply   Divide   **2 x 4 = ____ cups**

6. How does 1 cup change it?   **2 quarts 1 cup = ? cups**

Add 1 more cup.   2 x 4 = 8 + ____ = ____ cups

#2 1. What's the basic fact?   **1 gallon 3 qts = ? quarts**

Name 2 steps to find the answer.   **1 gallons = ____ quarts**

1 x 4 = ____ + ____ = ____ quarts

2. What fact changes cups to quarts?   **8 cups = ? quarts**

Do you multiply or divide to find it?   **1 quart = ____ cups**

Circle one   Multiply   Divide   **8 ÷ 4 = ____ quarts**

3. How do you change 9 cups?   **9 cups = ? quarts ? cups**

Subtract.   **8 ÷ 4 = ____ qts**   What is left?

9 cups = 2 quarts ____ cup

4. How does 11 cups change it?   **11 cups = ? quarts ? cups**

11 cups = 2 quarts ____ cup

#3 Change units with cups and quarts.   Calculator?
                                                             yes   no

1.   4 quarts = ____ cups   9 quarts = ____ cups
2.   7 quarts = ____ cups   11 quarts = ____ cups
3.   7 cups = ___ qts ___ cups   17 cups = ___ qts ___ cup
4.   1 qt 2 cups = ____ cups   4 qt 3 cups = ____ cups

#4 Change units with quarts and gallons.   Calculator?
                                                             yes   no

1.   6 gallons = ____ quarts   6 gallons = ____ quarts
2.   3 gal 2 qts = ____ qts   3 gals 2 qts = ____ qts
3.   21 qts = ___ gals ___ qts   21 qts = ___ gals ___ qts
4.   3 gals 2 qts = ____ qts   3 gals 2 quarts = ____ qts

Review 1. How many cups are in a can of 16 ounce pop?_____   Calculator?
2. A quart is how many cups? _____   3. A gallon is how many quarts? _____   yes   no
4. What number compares cups, quarts, and gallons? _____

Yes or correct?
5.   4 quarts = 12 cups   yes or ____ cups
6.   7 cups = 1 qt 3 cups   yes or ____ qts ___ cups
7.   1 qt 2 cups = 8 cups   yes or ____ cups
8.   6 gallons = 24 quarts   yes or ____ quarts
9.   3 gal 2 qts = 14 quarts   yes or ____ quarts
10. 21 qts = 4 gals 1 qt   yes or ____ gal ___ qts
11. 1 qt 2 cups = 8 cups   yes or ____ cups

12. Mr G's semi takes 10.5 gallons to change the oil. How many quarts is it? _____ qt
13. JJ can get 7 cups in his thermos. How many quarts/cups is it? _____ qt _____ cups
14. Keya's fish tank takes 17 qts. How many gallons/quarts is that? _____ gal _____ qt

Ch 3 Ls 2 How metrics measures liquid. 23

_____ #1 #2 ____/13 #3 #4 ____/12 R ___/ 9 Total ____/30 _____
            Name   Checker

#1 1. 1 gallon is about how many liters? _____

2. How do you compare a centiliter to what we drink? _____

3. A liter is how many centiliters? _____   4. 1 liter is how many deciliters? _____

5. What's the basic fact?   **3 liters = ? centiliters**

1 liter = ____ centiliters   3 liters = _____ centiliters

6. 2.8 liters is how many cl?   **2.8 liters = ? centiliters**

                                   2.8 liters = _____ centiliters

7. Change centiliters to liters.   **410 cl = ? liters**

                                   410 cl = _____ liters

#2 1. How many decilters are in a liter? _____

2. How much is a deciliter compared to what we drink? _____

3. How does the decimal move?   **1.6 liters = ? deciliters**

Move 1 place to the _____.   1.6 liters = ____ deciliters

4. Change to deciliters.   **3.5 liters = ? dl**

                                   3.5 liters = _____ dl

5. Change deciliters to liters.   **500 dl = ? liters**

                                   500 dl = _____ liters

6. Change deciliters to liters.   **60 dl = ? liters**

                                   60 dl = _____ liters

#3 Change liters with centiliters.   Calculator?
                                                       yes   no

1. 2 liters = _____ centiliters   3.7 liters = _____ centiliters
2. 2 liters 50 cl = _____ cl   3 liters 78 cl = _____ cl
3. 45 centiliters = _____ liters   570 centiliters = _____ liters
4. 78 centiliters = _____ liters   308 centiliters = _____ liters

#4 Change liters with deciliters.
                                                       Calculator?
                                                       yes   no

1. 2 liters = _____ deciliters   5 liters = _____ deciliters
2. 2 liters 3 dl = _____ dl   4 liters 5 dl = _____ dl
3. 65 deciliters = _____ liters   570 deciliters = _____ liters
4. 170 dl = _____ liters   213 dl = _____ liters

Review 1. 1 gallon is about how many liters? _____   Calculator?
2. How do you compare a centiliter to what we drink? _____   yes   no
3. How much is a deciliter compared to what we drink? _____
4. A liter is how many centiliters? _____   5. 1 liter is how many deciliters? _____

6. 2.8 liters = _____ centiliters   7. 1.6 liters = _____ centiliters
8. 3.5 liters = _____ deciliters   9. 4.2 liters = _____ deciliters
10. 570 centiliters = _____ liters   11. 680 centiliters = _____ liters
12. 45 deciliters = _____ liters   13. 32 deciliters = _____ liters
14. 2 liters 3 dl = _____ dl   15. 3 liters 6 dl = _____ dl
16. 4 liters 25 cl = _____ cl   17. 5 liters 50 cl = _____ cl

18. Ojas needs 3 liters of water for the camping trip.
    He has 11 liters. Is it enough?   Yes or _____

19. The shower uses 1 10th of a liter per minute. JT took
    a 12 minute shower. How many liters did it use?   Liters _____

Ch 3 Ls 3 Milliliters and Kilometers 25

_____ #1 #2 ____/ 11 #3 #4 ____/ 16 R ___/ 21 T ____/ 48 _____
         Name   Checker

#1 1. How many milliliters are in 1 liter? _____
    2. What do you compare a milliliter to? _____
    3. How many milliliters are in a centiliter? _____

       4. What's the basic fact?   **5.1 liters = ? milliliters**

         How does the decimal move?   1 liter = _____ milliliters

         Move the decimal right ___ places.   5.1 liters = _____ milliliters

       5. Change 0.7 liters to centiliters.   **0.7 liters = ? ml**

                                  0.7 liters = _____ ml

       6. Change 570 ml to liters.   **570 ml = ? liters**

                                  570 ml = _____ liters

#2 1. How many liters are in a kiloliter? _____
    2. What kind of problem would you use kiloliters with? _____

       3. What's the basic fact?   **2,800 liters = ? kiloliters**

         How does the decimal move?   1 kiloliter = _____ liters

         Move the decimal _____.   2,800 liters = _____ kiloliters

       4. Change 7 kiloiliters to gallons.   **7 kls = ? gallons**

                                  7 kls = _____ gallons

       5. Change 1.5 kiloliters to gallons.   **1.5 kl = ? gallons**

                                  1.5 kl = _____ gallons

#3 Change liters with milliliters. Calculator? yes no

1. 0.1 liters = _____ milliliters   2.0 liters = _____ milliliters
2. 2.8 liters = _____ milliliters   3.4 liters = _____ milliliters
3. 45 milliliters = _____ liters   570 milliliters = _____ liters
4. 780 milliliters = _____ liters   1200 milliliters = _____ liters

#4 Change liters with kiloliters. Calculator? yes no

1. 4 kiloliters = _____ liters   5.7 kiloliters = _____ liters
2. 2.6 kiloliters = _____ liters   7.5 kiloliters = _____ liters
3. 2300 L = _____ kl   4600 L = _____ kl
4. 5200 L = _____ kl   6800 L = _____ kl

Review Calculator? yes no

1. How many milliliters are in 1 liter? _____
2. What do you compare a milliliter to? _____
3. How many milliliters are in a centiliter? _____
4. How many liters are in a kiloliter? _____
5. What kind of problem would you use kiloliters with? _____

6. 0.2 liters = _____ milliliters   7. 0.8 liters = _____ milliliters
8. 1.8 liters = _____ milliliters   9. 3.5 liters = _____ milliliters
10. 45 milliliters = _____ liters   11. 57 milliliters = _____ liters
12. 4 kiloliters = _____ liters   13. 8 kiloliters = _____ liters
14. 5.7 kiloliters = _____ liters   15. 6.7 kiloliters = _____ liters
16. 1 kl 20 liters = _____ liters   17. 4 kl 500 liters = _____ liters
18. 3 kl 50 liters = _____ liters   19. 5 kl 125 liters = _____ liters

20. A juice has 1.2 centiliters guave. How many milliliters is that? _____ mL
21. Mrs K's pool takes 2.5 kiloliters water. How many gallons is that? _____ gal

Ch 3 Ls 4 Smaller units, Teaspoons   27

_____ #1 #2 ____/ 10 #3 #4 ____/ 22 R ___/ 16 T ____/ 48 _____
         Name   Checker

#1 1. How many teaspoons are in a centiliter? _____

   2. What's the basic fact?   **3 centiliters = ? teaspoons**

      What's the answer?   1 centiliter = ___ tsps

                                    3 centiliters = ____ tsps

   3. Change to teaspoons.   **3 cl = ? teaspoons**

                                    3 cl = ___ tsps

#2 1. How many teaspoons are in a tablespoon? _____

   2. What's the basic fact?   **4 tablespoons = ? teaspoons**

      What's the answer?   1 tablespoon = ___ tsps

                                    4 Tbsp = ____ tsps

   3. Change 2 tablespoons.   **2 tablespoons = ? tsps**

                                    2 Tbsp = ____ tsps

   4. How many teaspoons are in an ounce? _____
   5. How many teaspoons are in 1 centiliter, tablespoon and ounce? _____

   6. What's the basic fact?   **3 ounces = ? teaspoons**

      What's the answer?   1 ounce = ___ tsps

                                    3 ozs = ____ tsps

   7. Change 5 oz to teaspoons.   **5 oz = ? teaspoons**

                                    5 oz = ___ teaspoons

#3 Change teaspoons, tablespoons, and ounces.   Calculator? yes no

1. 3 oz = _____ tsps   5 oz = _____ tsps
2. 2 Tbsps = _____ teaspns   6 Tbsp = _____ tsps
3. 6 tsps = _____ Tablespns   12 tsps = _____ Tbsps
4. 18 tsps = _____ ounces   24 tsps = _____ ounces
5. 5 Tbsps = _____ teaspns   6 Tbsp = _____ tsps
6. 3 oz = _____ tsps   5 oz = _____ tsps

#4 Change teaspoons and centiliters.   Calculator? yes no

1. 4 tsps = _____ cl   10 tsps = _____ cl
2. 6 cl = _____ tsps   5 cl = _____ tsps
3. 2 oz = _____ tsps   4 oz = _____ tsps
4. 3 Tbps = _____ tsps   6 Tbps = _____ tsps
5. 12 Tbps = _____ tsps   15 Tbps = _____ tsps

Review
1. How many teaspoons are in a centiliter? _____   Calculator? yes no
2. How many teaspoons are in a tablespoon? _____
3. How many teaspoons are in an ounce? _____
4. How many teaspoons are in 1 centiliter, tablespoon and ounce? _____

5. 2 cl = _____ tsps   3 cl = _____ tsps
6. 5 cl = _____ tsps   6 cl = _____ tsps
7. 3 oz = _____ tsps   5 oz = _____ tsps
8. 2 Tbps = _____ teaspns   4 Tbps = _____ tsps
9. 6 tsps = _____ Tbsp   15 tsps = _____ Tbsp
10. 18 tsps = _____ ounces   24 tsps = _____ ounces

Review Problems 29

_____ #1 #2 #3 ____/37 #4 ____/12 Total ____/49
           Name

#1 1. Cup _____
    2. Quart _____
    3. Gallon _____
    4. Liter _____
    5. Centiliter _____
    6. Deciliter _____
    7. Millimeter _____
    8. Kiloliter _____
    9. Teaspoon _____

#2 Change these measurements.   Calculator?
                                                         yes   no
1.   4 qts 2 cups = ____ cups   7 qt 1 cups = ____ cups
2.   17 qts = ____ gals ____ qts   25 qts = ____ gals ____ qts
3.   7 gallons = ____ quarts   9 gallons = ____ quarts
4.   2 gals 3 qts = ____ qts   4 gals 2 qts = ____ qt
5.   7 gals 1 qts = ____ qts   8 gals 3 qts = ____ qts

#3 How long is each measurement?   Calculator?
                                                         yes   no
1.   1.7 liters = ____ centiliters   2 liters 80 cl = ____ cl
2.   4 liters 20 cl = ____ cl   67 centiliters = ____ liters
3.   75 centiliters = ____ liters   153 centiliters = ____ liters
4.   4 liters 5 dl = ____ dl   6 liters 8 dl = ____ dl
5.   78 deciliters = ____ liters   350 deciliters = ____ liters
6.   1.7 liters = _____ milliliters 2.8 liters = _____ milliliters
7.   740 milliliters = ____ liters   890 milliliters = ____ liters
8.   4 kiloliters = _____ liters   5.5 kiloliters = ____ liters
9.   4300 L = _____ kl   5600 L = _____ kl

#4 Story problems.

1. Ojas has 8 teaspoons of salt. How many centiliters is that? _____   Calculator? yes   no

2. Keya's baby weighs 7 lbs 14 oz. A doctor said the baby will gain about 4 oz per week. How much will it weigh in 2 months? _____

3. Juan wants 3 gallons of oil, but they only sell half quarts. How many will he need to get? _____

4. Eight ounces serving of horsey sauce has 35 milliliters of egg yolk. How many centiliters is that? _____

5. Mrs D goes to the grocery for 12 oz Pepper Sauce. It contains 72 milliliters of brown sugar. How many centiliters is it? _____

6. Mrs C has 7 tablespoons of flour, but the recipe wants it to be teaspoons. How much is it? _____

7. Mr B is looking for 2.5 kiloliter pool. He found one with 3.0 kL. What is the difference in gallons? _____

8. Once a day Mrs J waters the trees she raises. Each tree gets 2 cups of water. If Mrs J has 16 trees, how much water does she use in a week in gallons? _____

9. JJ bought 8 gallons of cider for a party. After the party there were still 9 quarts of punch left. How much cider was used at the party? _____

10. Mr W watched a TV show that said he's supposed to drink 10 cups of water each day. How many quarts of water is 10 cups? _____

11. Sana had four 2-liter bottles of water. She poured the water into 25 centiliter glasses. How many glasses did she fill before she ran out of water? _____

12. A vial of medicine contains 8 centiliters. One dose of the medicine is 4 milliliters. How many doses of medicine are in the vial? _____

You may take the quiz.

Ch 4 Ls 1 Ounces and Pounds   31

_____ #1 #2 ____ / 9   #3 #4 ____ / 20   R ___ / 11   T ____ / 30 _____
            Name   Checker

#1 1. How many ounces are in a pound? _____
    2. How many ounces are in a quarter of pound? _____
    3. How do you change 2 pounds to ounces? _____

        4. What's the basic fact?   **2 pounds = ? oz**

        What 2 steps finds the answer?   1 pound = ___ ounces

                                  Multiply 2 x 16 = ___ ounces

        5. Change 2 lb 3 ounces to ounces.   **2 lb 3 ounces = ? oz**

                                  ___ x 16 = ___ + ___ = ___ oz

        6. Change 3 lb 5 ounces to ounces.   **3 lb 5 ounces = ? oz**

                                  ___ x 16 = ___ + ___ = ___ oz

#2 1. Change 20 oz to pounds. Basic fact?   **20 ounces = ? pounds**

        20 ounces is how ounces?   1 pound = ___ ounces

        Whole fact = ___ x 16 = ___ lbs   Subtract ___ = ___ ounces

        2. What 2 steps finds the answer?   **40 ounces = ? lb ? oz**

        Whole fact = ___ x 16 = ___ lbs   Subtract ___ = ___ ounces

        3. What 2 steps finds the answer?   **50 ounces = ? lb ? oz**

        Whole fact = ___ x 16 = ___ lbs   Subtract ___ = ___ ounces

#3 Change with ounces and pounds.   Calculator? yes no

1. 2 pounds = ____ oz   3 pounds = ____ oz
2. 4 pounds = ____ oz   8 pounds = ____ oz
3. 32 ounces = ____ lb   48 ounces = ____ lb
4. 80 ounces = ____ lb   96 ounces = ____ lb
5. 128 ounces = ____ lb   54 ounces = ____ lb

#4 Change with ounces and pounds. Pt 2   Calculator? yes no

1. 1 pounds 3 oz = ____ oz   2 pounds 6 oz = _____ oz
2. 4 pounds 10 oz = ____ oz   5 pounds 12 oz = _____ oz
3. 28 ounces = ___ lb ___ oz   40 ounces = ___ lb ___ oz
4. 50 ounces = ___ lb ___ oz   70 ounces = ___ lb ___ oz
5. 60 ounces = ___ lb ___ oz   90 ounces = ___ lb ___ oz

Review 1. How many ounces are in a pound? _____   Calculator? yes no
2. How many ounces are in a quarter of pound? _____
3. How do you change 2 pounds to ounces? _____
4. What 2 steps do you use to change 20 oz to lb? _____

Yes or correct?

5. 2 pounds 8 oz = 38 oz   yes or ____ oz
6. 3 pounds 7 oz = 53 oz   yes or ____ oz
7. 4 pounds 10 oz = 74 oz   yes or ____ oz
8. 20 ounces = 1 lb 4 oz   yes or ____ lb ____ oz
9. 30 ounces = 1 lb 12 oz   yes or ____ lb ____ oz
10. 50 ounces = 3 lb 4 oz   yes or ____ lb ____ oz
11. 70 ounces = 4 lb 8 oz   yes or ____ lb ____ oz

Ch 4 Ls 2 Ton, Ton, Ton, Ton, Tonnnnn   33

_____ #1 #2 ____/ 9   #3 #4 ____/ 16   R ___/ 15   T ____/ 40 _____
              Name   Checker

#1 1. How many pounds are in 1 ton? _____   How many are in 3 tons? _____

2. Change to lbs. What are the 2 steps?   **2 tons 1,400 lbs = ? lb**

_____ + _____ = _____ lbs

3. How many pounds is 2 tons 400 lbs?   **3 tons 1,800 lbs = ? lb**

_____ + _____ = _____ lbs

4. Change 7,100 lbs to tons and pounds.   **7,100 lbs = ? tons ? lb**

6000 lb is ___ tons and _____ lbs

5. Change 3,500 lbs to tons.   **3,500 lbs = ? tons ? lb**

2000 lb is ___ tons and _____ lbs

#2 1. How many pounds is 1 10th of a ton? _____ pounds   3 10ths? _____ pounds

2. 2.3 tons is how many pounds?   **2.3 tons = ? pounds**

2 tons is 2000 x ___ = _____ lbs

Each 10th of a ton is 200 lb.   0.3 is ___ x 200 = _____ lbs

3. 1.7 tons is how many pounds?   **1.7 tons = ? pounds**

Add _____ + _____ is _____ lbs

4. 4.5 tons is how many pounds?   **4.5 tons = ? pounds**

Add _____ + _____ is _____ lbs

#3 Change with pounds and tons.   Calculator? yes no

1. 1 tons 300 lbs = _____ lb   2 tons 600 lbs = _____ lb
2. 2 tons 1400 lbs = _____ lb   4 tons 1200 lbs = _____ lb
3. 5600 lbs = ___ tons _____ lb   6400 lbs = ___ tons _____ lb
4. 2800 lbs = ___ tons _____ lb   4500 lbs = ___ tons _____ lb

#4 Change with pounds and 10ths of a ton.   Calculator? yes no

1. 1600 lbs = _____ tons   3200 lbs = _____ tons
2. 4200 lbs = _____ tons   7000 lbs = _____ tons
3. 1.3 tons = _____ lbs   2.5 tons = _____ lbs
4. 3.8 tons = _____ lbs   8.7 tons = _____ lbs

Review 1. How many pounds are in 1 ton? _____ How many are in 3 tons? _____   Calculator? yes no
2. How many pounds is 1 10th of a ton? _____ pounds   3 10ths? _____ pounds

Yes or correct.
4. 2 tons 1400 lbs = 4400 lb   yes or _____ lb
5. 3 tons 1800 lbs = 7800 lb   yes or _____ lb
6. 7400 lbs = 3 tons 400 lb   yes or _____ t _____ lb
7. 9500 lbs = 4 tons 500 lb   yes or _____ t _____ lb
8. 2.6 tons = 5400 pounds   yes or _____ lbs
9. 4.7 tons = 8400 pounds   yes or _____ lbs
10. 3400 pounds = 1.8 tons   yes or _____ tons
11. 9400 pounds = 4.5 tons   yes or _____ tons

12. A truck can carry 2.7 tons. How many pounds is that? _____
13. A semi truck is rated for 15 tons. How many pounds is it? _____

Ch 4 Ls 3 How metrics meausres weight.   35

_____ #1 #2 ____/ 13 #3 #4 ____/ 2 0 R ___/ 14 T ____/30 _____
         Name   Checker

#1 1. How many grams are in a kilogram? _____
    2. Name sonething that weighs 1 gram? _____
    3. How many milligrams are in a gram? _____   4. How are mg used? _____

    5. Change to kilograms.   **5000 grams = ? kg**

                          ___ x _____ = _____ kilograms

    6 Change to grams.   **2 kg 300 grams = ? grams**

                          _____ + _____ = _____ grams

    7. Change to milligrams.   **3 g 200 mg = ? mg**

                          3 g 200 mg = _____ mg

    8. Change to grams and mg.   **4200 mg = ? g ? mg**

                          4200 mg = ___ g _____ mg

#2 1. How much does 1 kilogram weigh in pounds? _____
    2. How much do 100 kilograms weigh in pounds? _____

       3. What's the basic fact?   **4 kilogram = ? pounds**

                          Multiply 2.2 x 4 = ____ pounds

       4. What's the basic fact?   **30 kilogram = ? pounds**

                          Multiply 2.2 x 30 = _____ pounds

       5. What's the basic fact?   **50 kilogram = ? pounds**

                          Multiply 2.2 x 50 = _____ pounds

#3 Change units with grams, centigrams, and kilograms. Calculator? yes no

1. 1 g 20 cg = _____ cg   3 g 60 cg = _____ cg
2. 2 g 500 mg = _____ mg   4 g 200 mg = _____ mg
3. 3600 mg = ___ g _____ mg 6500 mg = ___ g _____ mg
4. 240 cg = ___ g _____ cg   630 cg = ___ g _____ cg

#4 Change units with grams, kilograms, and pounds. Calculator? yes no

1. 4200 gs = _____ kg   8500 grams = _____ kg
2. 2 kg 700 gs = _____ g   3 kg 500 gs = _____ g
3. 5 kilograms = _____ lbs   7 kilograms = _____ lbs
4. 10 kilograms = _____ lbs   20 kilograms = _____ lbs
5. 4.4 pounds = _____ kgs   66 pounds = _____ kgs
6. 88 pounds = _____ kgs   110 pounds = _____ kgs

Review 1. How many grams are in a kilogram? _____ Calculator? yes no

2. Name sonething that weighs 1 gram? _____

3. How many milligrams are in a gram? _____   4. How are mg used? _____

5. How much does 1 kilogram weigh in pounds? _____

6. How much do 100 kilograms weigh in pounds? _____

Yes or correct?

7. 3 g 20 cg = 3200 cg   yes or _____ cg
8. 5 g 600 mg = 560 mg   yes or _____ mg
9. 4600 mg = 4 g 60 mg   yes or ___ g ___ mg
10. 570 cg = 5 g 70 cg   yes or ___ g ___ cg
11. 5030 grams = 5.3 kg   yes or _____ kg
12. 2 kg 300 grams = 230 g   yes or _____ g
13. 6 kilograms = 12.2 lbs   yes or _____ lbs
14. 50 pounds = 107 kgs   yes or _____ kg

Ch 4 Ls 4 Celsius and Fahrenheit 37

_____ #1 #2 ____/14 #3 #4 ____/12 R ___/ 9 Total ____/30 _____
              Name   Checker

#1 1. What is freezing and boiling point in fahrenheit?   Freezing _____   Boiling _____

2. What is freezing and boiling point in celsius?   Freezing _____   Boiling _____

3. How many fahrenheit degrees do you add for each 10 in celsius? _____

4. What are 10 and 20 in fahrenheit?   **10 Celsius**   **20 Celsius**

    _____ fahrenheit   _____ fahrenheit

5. What are 30 and 40 in fahrenheit?   **30 Celsius**   **40 Celsius**

    _____ fahrenheit   _____ fahrenheit

6. What are - 10 and - 20 in fahrenheit?   **- 10 Celsius**   **- 20 Celsius**

    _____ fahrenheit   _____ fahrenheit

7. What are 68 and 86 in Celsius?   **68 fahrenheit   86 fahrenheit**

    _____ Celsius   _____ Celsius

8. What are 4 and - 14 in Celsius?   **4 fahrenheit   - 14 fahrenheit**

    _____ Celsius   _____ Celsius

#2   Decide: winter coat, light coat, T shirt, shorts, or swim trunks?

1. The average person.   **30 Celsius**   **- 10 Celsius**

    _____   _____

2. Wears this outside.   **40 Celsius**   **10 Celsius**

    _____   _____

3. Think about seasons.   **- 18 fahrenheit**   **50 fahrenheit**

    _____   _____

#3 Change Fahrenheit with Celsius.   Calculator?
                                                                  yes   no

1. 10 C = ____ F   30 C = ____ F   - 10 C = ____ F
2. 20 C = ____ F   40 C = ____ F   - 20 C = ____ F
3. 50 F = ____ C   14 F = ____ C   68 F = ____ C
4. - 4 F = ____ C   86 F = ____ C   - 22 F = ____ C

#4 Decide: winter coat, light coat, t shirt, shorts, or swim trunks?   Calculator?
                                                                         yes   no

1. 10 C = _____   40 C = _____
2. 20 C = _____   - 10 C = _____
3. 86 F = _____   - 4 F = _____
4. 50 F = _____   104 F = _____

Review 1. What is freezing and boiling point in fahrenheit? Freezing _____ Boiling _____   Calculator?
        2. What is freezing and boiling point in celsius? Freezing _____ Boiling _____   yes   no
        3. How many fahrenheit degrees do you add for each 10 in celsius? _____

4. The recipe is to be at 300 F, but the oven is Celsius. What should
   the oven be set at in Celsius?   150 C   200 C   300 C

5. If the oven is set to be at 400 F, but the oven is Celsius. What should
   the oven be set at in Celsius?   150 C   200 C   300 C

6. If the is set to be at 600 F, but the oven is Celsius. What should
   the oven be set at in Celsius?   200 C   300 C   400 C

Change Fahrenheit with Celsius.

7. 10 C = ____ F   - 20 C = ____ F   40 C = ____ F
8. 30 C = ____ F   - 10 C = ____ F   - 22 F = ____ C
9. 68 F = ____ C   104 F = ____ C   50 F = ____ C

Review Problems

Name _____ #1 #2 #3 ____/ 38   #4 #5 ____/ 22   T____/ 60

#1
1. Pound _____
2. Ounce _____
3. Ton _____
4. Gram _____
5. Centigram _____
6. Milligram _____
7. Decigram _____
8. Kilogram _____
9. Celsius _____
10. Fahrenheit _____

#2 Change these measurements.   Calculator? yes   no

1. 3 pounds = ____ oz   6 pounds = ____ oz
2. 30 ounces = ____ lb ____ oz 45 ounces = ____ lb
3. 1 lb 6 oz = ____ oz   2 lbs 14 oz = _____ oz
4. 50 ounces = ___ lb ___ oz   70 ounces = ___ lb ___ oz
5. 4 tons 800 lbs = _____ lb   5 tons 1600 lbs = _____ lb
6. 5700 lbs = ____ tons   7200 lbs = ____ tons
7. 8500 lbs = ____ tons   9200 lbs = ____ tons
8. 1.7 tons = _____ lbs   3.8 tons = _____ lbs

#3 How long is each measurement?   Calculator? yes   no

1. 3 g 80 cg = _____ cg   4 g 20 cg = _____ cg
2. 6 g 500 mg = _____ mg   7 g 900 mg = _____ mg
3. 4600 gs = ____ kg   6300 grams = ____ kg
4. 2 kg 60 gs = _____ g   4 kg 3 gs = _____ g
5. 5 kilograms = _____ lbs   8 kilograms = _____ lbs
6. 6.6 pounds = ____ kgs   110 pounds = ____ kgs

#4 Change Fahrenheit with Celsius.   Calculator?
                                                                      yes   no

1.   - 4 F = _____ C   86 F = _____ C   - 22 F = _____ C

2.   10 C = _____ F   40 C = _____ F   - 20 C = _____ F

3.   20 C = _____ F   30 C = _____ F   - 10 C = _____ F

4.   50 F = _____ C   14 F = _____ C   68 F = _____ C

                                                                      Calculator?
                                                                      yes   no

#5 Story problems.

1. Ojas put out 5 dL of food for the chipmunks. They ate 1/2 the food. How many dL of food did the chipmunks eat? _____

2. Myra's baby weighs 6 lbs 10 oz. A doctor said the baby will gain about 5 oz per week. How much will it weigh in 4 months? _____

3. JJ gave a 2 kilogram box of chocolates with 64 pieces. What's the average weight for each piece of chocolate in ounces? (answer in centigrams) _____

4. The average weight of a baby at birth is 7 pounds. How many ounces is this? _____

5. A peach weighs 30 centigrams. You buy 12 of them. They sell for Rs 60 per kilogram. How much will they cost? _____

6. A box contains 8 bags of sugar which weighs 12 kiograms. What is the weight of each bag? _____

7. Naveed put out 7 cups of food for the rabbits. They ate 1/2 the food. How many ounces of food did the rabbits eat? _____

8. A Boeing 747 weighs about 876,000 pounds. How many tons is that? _____

9. The Statue of Liberty weighs 225 tons. How many pounds is this? _____

10. A country exported 107,000,000 lb of coal in 2011. How many tons is this? _____

Ch 5 Ls 1 Range and Close Range 41

_____ #1 #2 ____ / 8   #3 #4 ____ / 22   R ___ / 7   T ____ / 37 _____
Name   Checker

#1 1. How do you find the range of a group of numbers? _____

_____

2. What's the 1st step to find a range?   40   70   20   60

What happens next?   The highest number is ___ and lowest is ___.

Subtract them.   ____ - ____ = ____

3. Find the range.   80   10   120   90

____ - ____ = ____

4. Find the range.   210   240   300   150

____ - ____ = ____

#2 1. How does 10% Rule find a close range? _____
2. If a 30 has a + or - range of 3, what's the range? _____

3. Find a close range. What's 10% of 40?   40   47

How do you find if it's close?   10% of 40 is ___

Close   Not Close   40 + ___ = ___

4. Find a close range. What's 10% of 60?   57   60

How do you find if it's close?   10% of 60 is ___

Close   Not Close   60 - ___ = ___

#3 Find 10%, then find the upper and lower range.   Calculator? yes no

1. **40**   10% is ___ Add 40 + _4_ = _44_   Subtract 40 - _4_ = _36_
2. **70**   10% is ___ Add 70 + ___ = ___   Subtract 70 - ___ = ___
3. **120** 10% is ___ Add 120 + ___ = ___   Subtract 120 - ___ = ___
4. **150** 10% is ___ Add 150 + ___ = ___   Subtract 150 - ___ = ___
5. **200** 10% is ___ Add 200 + ___ = ___   Subtract 200 - ___ = ___
6. **500** 10% is ___ Add 500 + ___ = ___   Subtract 500 - ___ = ___

#4 Find the range for each group of numbers.   Calculator? yes no

1. 40 48 2 12 10 is ___   110 7 20 90 130 is ___
2. 40 70 20 60 10 is ___   50 80 19 110 200 is ___
3. 11 90 130 80 is ___   90 130 280 170 is ___
4. 350 70 130 50 is ___   500 120 210 300 is ___
5. 200 120 330 is ___   260 730 460 430 is ___

Review 1. How do you find the range of a group of numbers? _____   Calculator? yes no
_____

2. How does 10% Rule find a close range? _____
3. If the range of temperature is from - 3 C to 38 C, what is it? _____ C
4. The industrial oven has a min 150 and max 1800 C. What is the range? _____ C
5. The XJ Car's range is - 15 to 190 kph. What is the range?   _____ kph
6. The range of lightbulbs is 10 to 120 watts. What is the range?   _____ watts
7. The range of a menu is Rs 50 to Rs 795. What is the range?   _____ rupees

Ch 5 Ls 2 Find the Average 43

_____ #1 #2 ____ / 9 #3 #4 ____ / 9 R ___ / 9 Total ____ / 27 _____
　　　　Name　　　　　　　　　　　　　　　　　　　　　　　　　　　　　　　　　　　Checker

#1 1. Name 2 steps to find average. _____

2. What does it mean by "the number of numbers"? _____

3. What's the 1st step to find this average?   10   0   10   0   0

　　　　　　Add the numbers.   10 + 0 + 10 + 0 + 0 = ____   What's the average?

　　　　　　　　　　　Divide by _____   ___ ÷ 5 = ___

4. What's the range and average?   4   7   6   5   9

　　　　___ - ___ = ___ range   ___ ÷ ___ = ___ average

5. What's the range and average?   20   30   110   140

　　　　___ - ___ = ___ range   ___ ÷ ___ = ___ average

#2 Find the average of the range.

1. What's 1st to find the average of a range? _____

2. What's the 2nd step? _____

3. What's the range for these numbers?   10   30   90   50

　　　　What's the average?   ___ - ___ = ___ range

　　　　　　　　　　　　　　　　　___ ÷ ___ = ___ average

4. What's the range for these numbers? 100   140   80   60

　　　　What's the average?   ___ - ___ = ___ range

　　　　　　　　　　　　　　　　　___ ÷ ___ = ___ average

#3 Find the range and average of the numbers.   Calculator?
   yes   no

1. 4 10 6 4 1 5   range is _____   average of 6 num _____
2. 10 30 20 40 30   range is _____   average of 5 num _____
3. 80 70 60 10 20   range is _____   average of 5 num _____
4. 100 600 500 400   range is _____   average of 4 num _____
5. 8 7 9 5 8 5   range is _____   average of 6 num _____

#4 Find the range and average of the range.   Calculator?
   yes   no

1. 14 17 9 10 25   range ____ to ____ average of it _____
2. 20 30 60 40 20   range ____ to ____ average of it _____
3. 120 190 230 300   range ____ to ____ average of it _____
4. 50 60 130 200   range ____ to ____ average of it _____

Review 1. Name 2 steps to find average. _____   Calculator?
   yes   no
2. What does it mean by "the number of numbers"? _____
3. What's 1st to find the average of a range? _____
4. What's the 2nd step? _____

Find range and average.

5. 200 400 450 375   range ____ to ____ average of it _____
6. 150 200 210 240   range ____ to ____ average of it _____
7. 12 7 6 20 10   range is _____   average of 5 num _____
8. 20 16 25 34 40   range is _____   average of 5 num _____

9. Ojas has grades of 46, 60, 70, and 80. Find range _____. His average is _____
10. Zara has grades of 85, 90, 100, and 85. Range _____. Her average is _____

Ch 5 Ls 3 What a Median Finds 45

_____ #1 #2 ____/ 9 #3 #4 ____/ 8 R ___/ 21 T ____/38 _____
            Name   Checker

#1 1. Where do you find medians in the real world? _____

    2. How is median used in math? _____

    3. What's the 1st step to find a median?   8   3   20   15   12

    Put the numbers _____. Finish it.   ____ ____ ____ ____ ____

                                        The median is _____.

    4. Find the median.   6   7   13   15   18

                                        The median is _____.

    5. Find the median.   12   26   27   35   42

                                        The median is _____.

#2 1. How do you find the median for an even number of numbers? 3 steps_____

    _____

    2. How do you find the median
       with the number in between?   3   8   12   15

       Find the _____ number.   What number is between 8 and 12?

                                     _____ is between 8 and 12.

    3. What is the median in between?   4   10   12   14   19   20

                                     _____ is between 12 and 14.

    4. What are both numbers median?   2   9   14   15   21   27

                                     _____ and _____ are the medians.

#3 Find the median. Use the middle with even numbers. Calculator? yes no

1. 3 8 12 16 23 28   12 15 23 25 32 35

    ____ is between 12 and 16.   ____ is between 23 and 25.

2. 7 10 16 20 32 40   32 36 40 44 50 59

    ____ is between 16 and 20.   ____ is between 40 and 44.

#4 Find the median. Use both numbers. Calculator? yes no

1. 3 8 12 15 23 28   11 14 24 27 32 35

    Both numbers are ___ and ___.   Both numbers are ___ and ___.

2. 31 37 40 50   42 45 50 55 56 65

    Both numbers are ___ and ___.   Both numbers are ___ and ___.

Review 1. Where do you find medians in the real world? _____ Calculator? yes no
2. How is median used in math? _____
3. How do you find the median for an even number of numbers? 3 steps _____
   _____

4. 8 12 15 30 30   Median ____ Range ____ Average 5 num ____

5. 5 14 16 24 26   Median ____ Range ____ Average 5 num ____

6. 7 21 30 32 40   Median ____ Range ____ Average 5 num ____

7. 20 120 150 230   Median ____ Range ____ Average 4 range ____

8. 16 140 160 300   Median ____ Range ____ Average 4 range ____

9. 50 190 240 280   Median ____ Range ____ Average 4 range ____

Ch 5 Ls 4 How Mode is Different   47

_____ #1 #2 ____/ 8 #3 ____/ 6 R ___/ 10 T ____/ 24 _____
                Name   Checker

#1 1. What does mode show you? _____
2. What kind of problems find the mode? _____
3. Which one is more accurate, average or median? _____
4. What's the mode of a group of numbers?   5 4 7 5 8

There's more ___s than other numbers.   ___ is the mode

#2 Find range and average, then mode and median.

1. Find the range and average.   40 70 60 40 10

Find mode and median.   range _____   average ___ ÷ ___ = ___

mode _____   median _____

2. Find range and average.   50 20 50 40 90

Find mode and median.   range _____   average ___ ÷ ___ = ___

mode _____   median _____

3. Find range and average.   120 40 60 80 120

Find mode and median.   range _____   average ___ ÷ ___ = ___

mode _____   median _____

4. Find range and average.   40 70 60 40 40

Find mode and median.   range _____   average ___ ÷ ___ = ___

mode _____   median _____

#3 Find the range, mode, and median.   Calculator? yes no

1. 3 4 4 8 15 16   Range ____ Mode ____ Median ____
2. 5 12 12 20 21   Range ____ Mode ____ Median ____
3. 18 18 25 30 35   Range ____ Mode ____ Median ____
4. 120 150 210 210   Range ____ Mode ____ Median ____
5. 140 160 160 250   Range ____ Mode ____ Median ____
6. 220 300 330 330   Range ____ Mode ____ Median ____

Review

1. What does mode show you? _____
2. What kind of problems find the mode? _____
3. Which one is more accurate, average or median? _____

Calculator? yes no

Story Problems

4. A local restaurant has prices from Rs 65 to Rs 840. What are the range and average?   range ____   average ____

5. Mitul has to read from pg 35 to 115 in 5 days (1 week). He wonders what the range and average of the pages is?   range ____   average ____

6. Mitul didn't read anything the 1st day. Now he has to read from pg 35 to 115 in 4 days. What's the average and range?   range ____   average ____

7. Mr D drove these miles. 123 375 450 240 He had to find average and the range for the business report.   range ____   average ____

8. Mr G has to keep the average of bikes he has for sale. It's 26 40 36 21 and 17. Find the average and range.   Average ____   Range ____

9. Ojas has grades of 60 60 75 85 and 90. Now he has to find the average and range.   Average ____   Range ____

10. Mr P wants the average and range for hamburgers sold by the hour. It's Rs 320 Rs 410 Rs 460 Rs 500 and Rs 360.   Average ____   Range ____

Review Problems 49

_____ #1 #2 #3 ____/ 21 #4 ____/ 12 Total ____/ 33
                Name

#1 1. Range _____
    2. Close Range _____
    3. Average _____
    4. Median _____
    5. Mode _____

#2 Find the range and average of the numbers.   Calculator?
                                                               yes   no

1. 5   30   6   40   4   8   range is _____ average of 6 num _____
2. 80 50   60 40   20   range is _____ average of 5 num _____
3. 200 700 600 300   range is _____ average of 4 num _____
4. 400 900 500 600   range is _____ average of 4 num _____
5. 9   7   14 5   3   5   range is _____ average of 6 num _____
6. 300 1200 500 800   range is _____ average of 4 num _____
7. 90 60   150 80   130   range is _____ average of 5 num _____
8. 16 8   5   20   18 30   range is _____ average of 6 num _____

#3 Find the median and mode. Use the middle with even numbers.   Calculator?
                                                                   yes   no

1. 22 46   70 44   46 60   mode is _____ median is _____
2. 13 35   27 27   32 35   mode is _____ median is _____
3. 200 600 200 400 80   mode is _____ median is _____
4. 31 37   40 31   50 120   mode is _____ median is _____
5. 400 600 500 600 700   mode is _____ median is _____
6. 40 12   12 16   23 30   mode is _____ median is _____
7. 30 15   35 20   32 35   mode is _____ median is _____
8. 32 20   16 20   32 50   mode is _____ median is _____

#5 Story problems.   Calculator?
                                                              yes   no

Two film reviewers posted the movies they watched in a year. Answer the qs.

	Jan.	Feb.	Mar.	Apr.	May	Jun	Jul	Aug	Sept.	Oct.	Nov.	Dec.
Amav	1	3	2	5	2	3	1	4	2	3	2	1
Zara	1	2	1	1	1	3	3	2	2	6	1	1

1. Comparing modes, which person went to the movies the least per month?   _____

2. Comparing medians, which person went to the movies the most per month?   _____

3. How many did movies did Amav average?   _____

4. How many did Zara average per month?   _____

5. A professional player played 6 games of basketball last week. He scored 25, 32, 25, 39, 23, and 25 points. What's the mode of his scores?   _____

6. The prices for a gallon of milk at four stores are listed below. Rs 45, Rs 50, Rs 60, and Rs 85. What is the range for the 4 stores?   _____

7. An ice cream shop records the number of ice creams sold per day as 60, 58, 82, 55, and 70. What's the average and range?   Average _____ Range _____

8. On a cold winter day, the temperature for 7 cities is recorded in Celsius. What is the average and range? -8 C, 2, -3, 4, 12, 0, and 9.   Average _____ Range _____

9. Anvi has grades of 82, 67, and 75 on 3 math tests. What grade does she need on her next test to have an average of exactly 80 for the four tests?   _____

10. 6 daily temperatures for Anckorage are listed below. - 13, 4, 2, -15, and 5 Celsius. What does the 6th day need to be to average 0?   _____

11. Mitul played 9 games on a basketball team, He averaging 18 points per game. How many points has Mitul scored?   _____

12. Five earthquakes and their magnitudes are listed. What is the average? 7.0, 6.2, 7.7, 6.7, and 6.4   _____

Ch 6 Ls 1   Bar and Picture Graphs   51

_____ #1 #2 ____/ 10 #3 #4 ____/ 8   R ___/ 5 T ____/ 23 _____
Name   Checker

#1 1. Name 2 things bar graphs are good at. _____
     2. How are picture graphs different from bar graphs? _____
     3. Name 3 reasons to use picture graphs. _____
     4. What is 1 reason to use bar graphs over picture graphs? _____

#2 Find the kpg for each car.

   1. What is the range between car 1 and car 2?

Highway MPG
Car 1 [==================] 24
Car 2 [=============] 18
0   5   10   15   20   25   30

Car 1 mpg ____   Car 2 mpg ____   ____ - ____ = ____ range

2. What is the average between them?

____ ÷ ____ = ____ average

Find the kpg for each truck.

Highway MPG
Truck 1 ☆☆☆ 15
Truck 2 ☆☆ ʒ 11
5   10   15   20   25

3. What is the range between truck 1 and truck 2?

Truck 1 mpg ____   Truck 2 mpg ____   ____ - ____ = ____ range

4. What is the average between them?

____ ÷ ____ = ____ average

Opinion

5. Which kind of graph is easier to read? _____
6. Which graph is more accurate? _____

#3 Answer the questions with the bar graph.   Calculator? yes no

Average Monthly Rainfall

1. What's the range for the graph?

____ - ____ = ____
              range

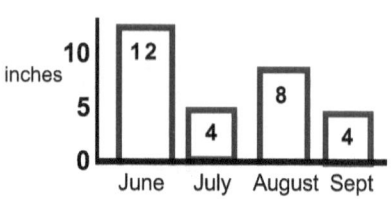

2. What is the median for the 2 months? _____

3. Is their a mode for the scores? _____

4. What is the average monthly rainfall? _____

Age of the Kids   Calculator? yes no

#4 1. What is the difference between Ojas and Zara?

____ - ____ = ____
              range

Amav 9
Ojas 17
TJ   15
Zara 7

0   5   10   15   20

2. What is the average for the 2 kids? _____

3. What is the median for the scores? _____

4. What is the average age of the kids? _____

Review 1. Name 2 things bar graphs are good at. _____

2. How are picture graphs different from bar graphs? _____

3. Name 3 reasons to use picture graphs. _____

4. What is 1 reason to use bar graphs over picture graphs? _____

_____

5. Would you rather use a picto or bar graph? _____

Ch 6 Ls 2   How a Graph Shows Time   53

_____ #1 #2 ____ / 5   #3 ____ / 10   R ____ / 3   T ____ / 18   _____
            Name   Checker

#1 1. How does a graph show time? _____

2. How do you follow time on a line graph? _____

3. How does a double graph show time differently? _____

#2 **Speed of cars**

1. Find the mph for car #1.

    How fast does the car
    speed up in 3 seconds?

Car 1 ——

Kph (graph 0–60 vs seconds 0–7)

It's going _____ kph at 3 seconds.   How fast is the car going at 5 seconds?

It's going _____ kph at 5 sec.   How much did it speed up
                                         between 3 and 5 seconds?

____ - ____ = _____ kph faster

Speed of trucks

2. Find the mpg for each truck.

    How fast does the car
    speed up in 2 seconds?

Truck 1 ——

Kph (graph 0–60 vs seconds 0–7)

It's going _____ kph at 2 sec starting at 20 mph.   How fast is the truck
                                                    going at 7 seconds?

It's going _____ kph at 7 sec.   How much did it speed up
                                         between 2 and 7 seconds?

____ - ____ = _____ kph faster
7 sec 2 sec

#3 Answer these line graph questions.   Calculator?
                                                                          yes   no

Fruit Basket Prices

What is the difference between
strawberry and apple baskets?

Apples ———

Strawberry ....

Total Rs 30 / Rs 25 / Rs 20 / Rs 15 / Rs 10 / Rs 5 / 0
Years 90 91 92 93 94 95 96 97

1. What do apples and straberries start out at in 1990? apples ____ strawberries ____

2. Compare their costs in 1990. ____ - ____ = ____   _____ is Rs ____ greater.

3. At what point have the baskets equaled out in price? _____

4. In 1997, are apples or strawberries more valuable? _____

5. Think of 1 reason for their change? _____

Indian Taxes Paid in 1959 - 63

Millions 50 / 40 / 30 / 20 / 10 / 0
27   33   40   45   45
Year 1959 1960 1961 1962 1963

6. What is the war that occurs during 1859 - 1863? _____

7. What is the range of millions raised? _____

8. What is the mode for this graph? _____

9. What is the average for the graph? _____

10. What is the median of the graph? _____

Review 1. How does a graph show time? _____   Calculator?
                                                            yes   no
2. How do you follow time on a line graph? _____

3. How does a double graph show time differently? _____

Ch 6 Ls 3 How Frequency Graphs Count Things/Histogram 55

_____ #1 #2 ____/ 11 #3 #4 ____/ 12 R ____/ 5 T ____/ 28 _____
              Name   Checker

#1 1. What does a frequency graph measure? _____
2. How does a frequency graph write a 5? _____
3. How is a stem and leaf graph different? _____
4. How does a stem graph count this? 9 0   2 3 7 8 _____
5. Define a histogram. _____
_____

#2 1. Count each of these.   卌 卌 | 卌 卌 卌 卌 卌

                           _____   _____

                                          Favorite Pie
2. How many like each pie?   Apple | 卌 卌 |||
                                 Chocolate| 卌 卌 卌 卌 |
                                 Peach | 卌 卌 卌 ||

Apple _____   Chocolate _____   Peach _____   What is the range?

3. Chocolate _____ - Apple _____ = _____ range   What is the median?

4. _____ is in the middle. What's the average for all 3?

        _____ + _____ = _____   _____ ÷ _____ = _____ average

6. Is the difference bigger between apple and peach or peach and chocolate?

      Peach _____ - Apple _____ = _____ difference   Chocolate _____ - Peach _____ = _____

7. How many people tasted the pies in all?

      Apple _____ + Chocolate _____ + Peach _____ = _____ total

#3 How many are in each?   Calculator? yes no

1. ||||| ||||| ||| _____ ||||| ||||| ||||| ||||| ||| _____
2. ||||| ||||| | _____ ||||| ||||| ||||| ||||| ||||| ||| _____
3. ||||| ||||| ||||| _____ ||||| ||||| ||||| ||||| ||||| _____

#4 Solve the Story Problems   Calculator? yes no

Grades for Math
Before the Final Exam

90s	0 0 1 4 8 8
80s	0 0 0 1 1 4 5 8 9
70s	0 0 1 1 3 5 8
60s	0 5 5

1. Which is the mode grade before the exam? _____

2. What is the range of grades? _____

3. How many total grades are there? _____

Grades for Math
After the Final Exam

90s	0 0 0 0 1 4 8 9
80s	0 0 0 1 1 1 4 5 8
70s	0 0 1 3 5 8
60s	5 8

4. Which is the mode grade after the exam? _____

5. What is the range of grades? _____

6. How did the grades change? _____

Review

1. What does a frequency graph measure? _____   Calculator? yes no
2. How does a frequency graph write a 5? _____
3. How is a stem and leaf graph different? _____
4. How does a stem graph count this? **90**   2 3 7 8   _____
5. Define a histogram. _____

Ch 6 Ls 4 Make a Survey and Find Percents   57

_____ #1 #2 ____ / 6   #3 ____ / 7   R ___ / 4   Total ____ / 17 _____
               Name   Checker

#1 1. How does a survey gather information? _____
2. Name 5 things to make a survey up. _____
_____
3. Name 3 steps to find a percent. _____
4. What number do the percents add to? _____

                              Boys   Girls
#2 1. How many boys and girls? |||| |||| | |||| |||| ||||

Make a percent of the total for each.
What's the 1st step?   ___ boys   ___ girls

What's the 2nd step for percents?   ____ + ____ = ____ total

How does it make percents?   Boys —   Girls —
                                    25   25

                              ____% are boys   ____% are girls

                              Boys   Girls
2. How many boys and girls are
     there? |||| ||| |||| |||| ||

Make a percent of the total for each.
What's the 1st step?   ___ boys   ___ girls

What's the 2nd step for percents?   ____ + ____ = ____ total

How does it make percents?   Boys —   Girls —
                                    20   20

                              ____% are boys   ____% are girls

#3 Solve these problems.   Calculator?
                                                              yes   no

                    Boys   Girls
                    ||||| ||||| | ||||| ||||| ||||| |||

1. Make a fraction for boys, then girls.   _____   _____

   Make a percent for each.   _____   _____

                    Boys   Girls
                    ||||| ||||| ||||| ||||| ||||| ||||| |||

2. Make a fraction for boys and girls.   _____   _____

   Make a percent for each.   _____   _____

Hours Ralph worked the 1st 4 weeks

Week 1 — 37
Week 2 — 34
Week 3 — 24
Week 4 — 25

(scale: 0 10 20 30 40)

3. What is the range of hours he works? _____

4. What is the mode for his work schedule? _____

5. What is the average of hours he worked? _____

6. He makes Rs 100 an hour. Compare the weeks. 37 x 100 = _____ 34 x 100 = _____

   24 x 100 = _____   25 x 100 = _____

7. What is the range of pay checks he received? _____

Review

1. How does a survey gather information? _____

2. Name 5 things to make a survey up. _____
   _____

3. Name 3 steps to find a percent. _____

4. What number do the percents add to? _____

Review Problems 59

_____ #1 #2 ____ /22 #3 ____ / 22 Total ____ / 44
Name

#1 1. Bar Graph _____
    2. Picture Graph _____
    3. Time Graph _____
    4. Double Graph _____
    5. Frequency Graph _____
    6. Stem and Leaf Graph _____
    7. Survey _____

#2 Story Problems   Calculator?
                                                              yes   no

1. Summer Camp Signups
   Pottery: 9
   Swim: 17
   Archery: 15
   Bugs: 7

   1. What is the range? _____
   2. How many chose bugs? _____
   3. How many chose swimming? _____
   4. How many total campers are there? _____
   5. What's the average? _____

2. Going Fishing
   Sana: 🐟🐟🐟🐟
   Hiran: 🐟🐟
   Pari: 🐟🐟🐟🐟🐟
   Mitul: 🐟🐟🐟

   6. What is the range? _____
   7. How many did Pari have? _____
   8. How many did Hiran have? _____
   9. How many fish did they catch in all? _____
   10. What's the average? _____

3. Soccer Goals
   Amav ········
   Ojas ─────

   11. What is the range? _____
   12. How many total hits for Amav? _____
   13. How many total hits for Ojas? _____
   14. What's the average? _____
   15. What's the mode? _____

Calculator? yes no

1. Cookies Sold

Adah	⧸⧹⧹⧹⧹ ‖
Eva	⧸⧹⧹⧹⧹ ⧸⧹⧹⧹⧹ ⧸⧹⧹⧹⧹ ⧸⧹⧹⧹⧹ ∣
Ira	⧸⧹⧹⧹⧹ ‖‖
Zana	⧸⧹⧹⧹⧹ ⧸⧹⧹⧹⧹ ⧸⧹⧹⧹⧹ ‖‖

Cookies Sold

1. What is the range? _____
2. How many did Adah sell? _____
3. How many did Ira sell? _____
4. How many total cookies were sold? _____
5. What's the average? _____

2. Final Test Grades

100s	0 2 6
90s	0 0 0 2 2 6 6 8
80s	0 0 0 0 2 4 6 8 8
70s	0 0 2 6 8
60s	0 2 8

6. What is the range? _____
7. What's the mode? _____
8. How many total students are there? _____
9. What's the top grade? _____

3. Favourite Colors

Yellow	9
Red	18
Green	14
Blue	12

0 5 10 15 20

10. What is the range? _____
11. How many chose green? _____
12. How many chose yellow? _____
13. What's the mode? _____
14. How many total students are there? _____

4. Pie Sales

Mitul	XXX
Sana	XXXXXXXX
Ira	XXXXX

Each X is 2 Pies

15. What is the range? _____
16. How many did Kim sell? _____
17. How many did Matt sell? _____
18. How many did Sonya sell? _____
19. What's the average? _____

5. Movie Tickets Sold

Week 1	⧸⧹⧹⧹⧹ ‖‖
Week 2	⧸⧹⧹⧹⧹ ⧸⧹⧹⧹⧹ ⧸⧹⧹⧹⧹ ‖
Week 3	⧸⧹⧹⧹⧹ ⧸⧹⧹⧹⧹ ‖

Each | is 5 tickets sold

20. What is the range? _____
21. Which week had the fewest? _____
22. What's the average? _____

Ch 7 Ls 1 Circle Graphs Use Circles   61

_____ #1 #2 ____/ 8 #3 ____/ 4 R ____/ 6 Total ____/ 18 _____
                    Name   Checker

#1 1. What percent do circle graphs add upto? _____

2. What do circle graphs use for information? _____

3. Name 3 steps to make percents of a number. _____

#2 1. What 3 percents add to get this circle?

_____% + _____% + _____% = 100%

2. Estimate 3 percents that add to 100%.

_____% + _____% + _____% = 100%

3. What 3 percents add to get this circle?

_____% + _____% + _____% = 100%

4. Estimate 3 percents that add to 100%.

_____% + _____% + _____% = 100%

5. Estimate 3 percents that add to 100%.

_____% + _____% + _____% = 100%

#3 Estimate the percents each circle adds.   Calculator?
                                                    yes   no

1.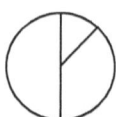

____% + ____% + ____% = 100%   ____% + ____% + ____% = 100%

2.

____% + ____% + ____% = 100%   ____% + ____% + ____% = 100%

Review 1. What percent do circle graphs add upto? _____   Calculator?
                                                                              yes   no
2. What do circle graphs use for information? _____
3. Name 2 steps to make percents from numbers. _____
_____

4. A store sold 70 diet pops, 420 regular pops, and 210 seltzer pop over the weekend. Make a circular graph.

____% + ____% + ____% = 100%

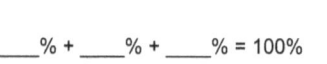
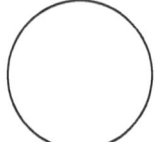

5. An auto store sold 18 trucks, 30 cars, and 12 suv's this week. Use a circular graph to show their results.

____% + ____% + ____% = 100%

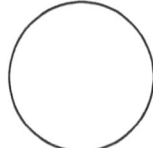

6. Twelve people at the restaurant prefer cricket, 26 like baskeball, and 37 prefer futball. Make a circular graph. (Estimate)

____% + ____% + ____% = 100%

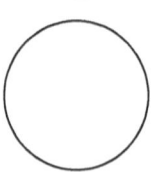

Ch 7 Ls 2   Venn Diagrams   63

_____ #1 #2 ____/ 12 #3 ____/ 5 R ___/ 4 Total ____/ 21 _____
          Name   Checker

#1 1. What do venn diagrams start with to make graphs? _____
      2. How does it show when both sets have same numbers? _____
      3. How do you write the union of both sets? _____
      4. What is the opposite of union? _____

#2 1. Make a union for   Set A (2 4 6 8)   Set B (7 8 9)
         sets A and B.

     2. What is the intersection
        of A and B?   _____

     3. Make a union for   Set A (1 2 3 4 5)   Set B (4 5 6)
         sets A and B.

     4. What is the intersection of A and B? _____

     5. Make a union for   Set A (10 20 30)   Set B (10 30)
         sets A and B.

     6. What is the intersection of A and B? _____

     7. Make a union for   Set A (15 30 45)   Set B (30 50)
         sets A and B.

     8. What is the intersection of A and B? _____

#3 Solve the Venn Diagrams story problems.   Calculator?
                                                                yes   no

1. JT surveyed 40 people to see if they preferred beef
   or chicken. 28 liked beef and 12 preferred chicken,
   but he had 7 that liked both beef and chicken. Make
   a Venn Diagram.

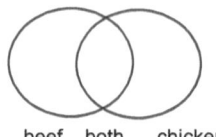
beef   both   chicken

2. Keya surveyed 75 people whether they had wood or
   carpet on their floors. 32 had just wood and 43 had
   carpet, but she found 17 of the people had both.
   Make a Venn Diagram for it.

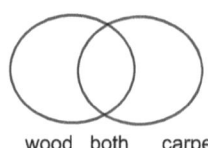
wood   both   carpet

3. Mrs K had 28 chocolate bars for her class. She
   polled and found 17 milk and 11 preferred dark
   chocolate. She then found 5 students liiked both
   of them. She made a Venn Diagram to show them.

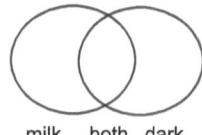
milk   both   dark

4. The principal asked 70 students how they got to
   school. 31 walked and 39 took the bus, but 14 of
   said they only took the bus when the weather was
   good. Make a Venn Diagram for him.

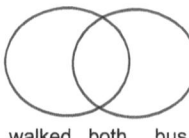
walked   both   bus

5. Mr J surveyed parents at the park and found 19
   had a boy and 22 had a girl. 12 of them had both
   boys and girls. Make a Venn Diagram.

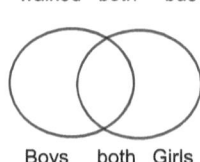
Boys   both   Girls

Review 1. What do venn diagrams start with to make graphs? _____

        2. How does it show when both sets have same numbers? _____

        3. How do you write the union of both sets? _____

        4. What is the opposite of union? _____

Ch 7 Ls 3 How Number Grids Show Information 65

_____ #1 #2 ____/ 7 #3 ____/ 8 R ____/ 5 Total ____/ 24 _____
　　　　Name　　　　　　　　　　　　　　　　　　　　　　　　　　　　　　　　　　　Checker

#1 1. What does a number grid for graphs look like? _____

2. How does number grid find a point? _____

3. How did these grids show correlations? _____

_____

4. Make a graph with these wins and losses.
What does the graph look like?

	#1	#2	#3	#4
Wins	2	3	5	4
Losses	4	3	1	2

**Wins and Losses by the
2014 Basketball team**

Wins (grid 0–6 vs Lost 0–7)

How did these grids show correlations?

#2 1. What does a key show on a map? _____

2. What does a map use to show distances on a map? _____

3. Make a graph with these cities.
What does the graph look like?

	A	B	C	D
Xs	2	4	6	7
Ys	6	3	3	2

**Position of 4 Cities
in the Greater Area**

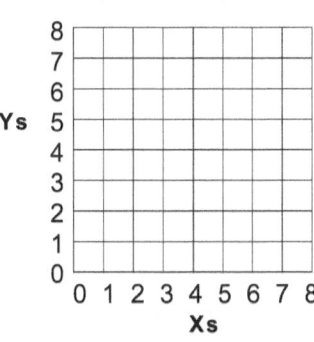

#3 Use a number grid for these problems.

Calculator? yes no

1. Make a graph with wins and losses.

	#1	#2	#3	#4	#5	#6
Wins	2	3	5	8	5	1
Losses	6	5	3	0	3	7

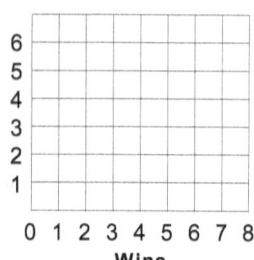

2. What is the range for wins by the soccer clubs? _____

3. What was the average number of losses? _____

4. What is the median for wins? _____

5. Make a graph with percent of free throws made.

	#1	#2	#3	#4
Percent	25	35	40	50

	#5	#6	#7	#8
Percent	55	45	35	40

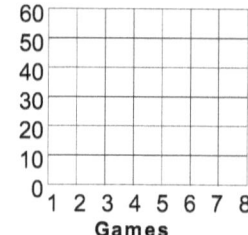

6. What is the range of free throws made? _____

7. What is the average for 8 games? _____

8. What is the mode for free throws made? _____

Review

1. What does a number grid for graphs look like? _____

2. How does number grid find a point? _____

3. How did these grids show correlations? _____

_____

4. What does a key show on a map? _____

5. What does a map use to show distances on a map? _____

Ch 7 Ls 4 Steps to Make a Graph   67

_____ #1 #2 ____/ 9 #3 ____/ 10 R ___/ 6 Total ____/ 25 _____
Name   Checker

#1 1. What are the 1st 2 steps to make a graph? _____
    2. Name 4 ways to collect information. _____
    _____
    3. What's the last step before making the graph? _____
    4. Name 3 things you need to make a graph. _____
    5. How do you find range for a scale? _____
    6. How does a large scale change the gaps? _____

	Dec	Jan	Feb	Mar
Alta	8	12	19	14
Cable	3	10	23	26

#2 1. Make a graph with the monthly rain totals in cm. What kind of graph would work best?

Use a Double Bar Graph or double line graph.

2. What scale would fit this graph?

Put _____ on the bottom
and _____ on the side.

centimeters
30
25
20
15
10
5
0
    months Dec Jan Feb Mar

3 TJ checked everywhere for loose change. Look at his data sheet below. Then use the table on the right to organize it.

Pesto 3   1 Rupee
5 Rupee   10 Rupee

What kind of graph will you make for it?

Bar Picture Time Frequency Stem/Leaf
Survey Circle Venn Number Grid

Make the graph!!

#3 Steps to Make a Graph   Calculator?
                                                              yes   no

Roll 2 die 25 times. Here are the results.

2	3	4	5	6	7	8	9	10	11	12
I	I	0	III	ꟼꟼ	III	III	ꟼꟼ	II	I	I

1. What was the range? _____ 2. What was the mode? _____

Do it yourself. What are results?

2	3	4	5	6	7	8	9	10	11	12

3. What was the range? _____ 4. What was the mode? _____

Daily High and Low Temperatures for Delhi, India.

Answer the questions on this graph.

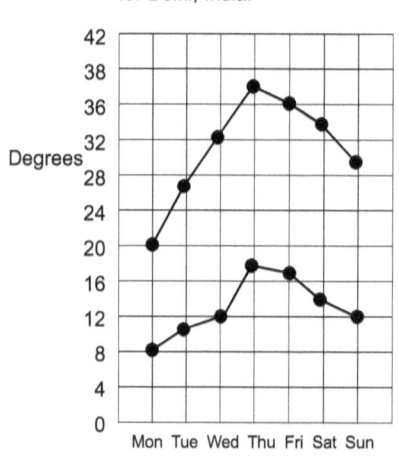

5. What is the highest temp? _____
6. What is the lowest temp? _____
7. What is the range? _____
8. What is the mode? _____
9. How many times was the high temperature at or above 35? _____
10. How many times was the low temperature at 15 or below? _____

Review 1. What are the 1st 2 steps to make a graph? _____

2. Name 4 ways to collect information. _____
_____

3. What's the last step before making the graph? _____
4. Name 3 things you need to make a graph. _____
5. How do you find range for a scale? _____
6. How does a large scale change the gaps? _____

Review Problems 69

_____ #1 #2 ____/ 19 #3 ____/ 20 Total ____/ 40
                Name

#1 1. **Circle Graph** _____

2. **Venn Diagram** _____

3. **Number Grid** _____

4. **Map** _____

5. **Scale** _____

#2 Story Problems   Calculator?
                                              yes   no

Pizza

1. 20 students voted for their favorite school pizza.

$\frac{1}{4}$ = sausage $\frac{1}{2}$ = cheese $\frac{1}{4}$ = pepperoni

1. How many picked cheese? _____
2. How many chose pepperoni? _____
3. How many chosse sauage? _____
4. What's the mode? _____
5. Which would you choose? _____

2. **Parents at the Park**

   12 (7) 15
   Boys both Girls

   JJ asked the parents how many of each they had there.

6. How many had boys? _____
7. How many had girls? _____
8. How many had both? _____
9. What's the problem with this graph? _____
   _____

3. **Favourite Colors**

White	4
Black	6
Green	3
Red	8
Blue	6
Yellow	7
Brown	2
Purple	5

The art teacher asked a class their favorite color.

10. What's the most favorite color? _____
11. What's the least favorite? _____
12. What is the range? _____
13. What's the mode? _____
14. How many total students are there? _____

#3 1. **Fish That Got Caught**

smallmouth bass 25
largemouth bass 12
northern pike 9
walleye 3

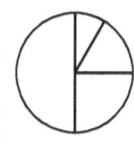

1. What is the range? _____
2. Which is the most caught? _____
3. Which is the least caught? _____
4. How many were caught in all? _____
5. What's the average? _____

Calculator?
yes   no

2. **Fish Vs Chicken**

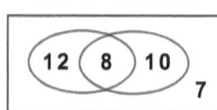

Fish Both Chicken

Sana asked 37 4th grade students
Fish vs Chicken at Arrow School, India.

6. What is the range? _____
7. How many chose Fish? _____
8. How many chose Chicken? _____
9. How many chose Both? _____
10. How many chose none? _____

3.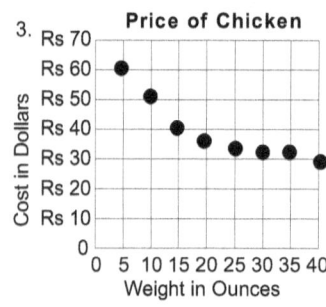

11. Which is the cheapest? _____
12. Which is the most expensive? _____
13. What is the range? _____
14. What's the mode? _____

4. **Treasure Map**

10
9
8
7
6
5
4
3
2
1
0
 0 1 2 3 4 5 6 7 8

Old River

15. Where is the mountain range? _____
16. Where is 1st part of the river? _____
17. Where is the 2nd part of the river? _____
18. Where is the cabin at? _____
19. Where is treasure the at? _____
20. What does the scale tell about the map? ____
_____

Each block is 5 10ths of a kilometer.

Ch 8 Ls 1 Words that Show Probability   71

_____ #1 #2 ____ /12 #3 ____/ 3 R ___/ 12 Total ____/ 27 _____
Name   Checker

#1 1. Name 2 steps to find a prediction. _____
2. What word tells about an event that can't possibly happen? _____
3. How about something not impossible but doesn't make sense? _____
4. Name 2 words people use for something that should happen. _____
_____
5. Name 2 words for something is 50/50. _____
6. What's a word for something that has to happen? _____

#2 1. A word for the chances.   **Ice melting in a 200 degree oven.**

_____

2. Think about this word.   **Flipping a coin and getting heads**

_____

3. What word is it?   **Getting wet when it's raining.**

_____

4. What word finds the chance? **Lightning during a rain storm**

_____

5. What word is it?   **Finding a coin on the ground**

_____

6. What word is it?   **Pick 7 of hearts from a card deck.**

_____

#3 What word describes the chances of this happening?   Calculator?
                                                            yes   no

1. A word for the chances.   **Finding a needle in a haystack.**

_____

2. Think about this word.   **Flipping a coin and getting tails.**

_____

3. What word is it?   **Staying dry outside during a rainstorm without an umbella.**

_____

Review   1. Name 2 steps to find a prediction. _____   Calculator?
2. What word tells about an event that can't possibly happen? _____   yes   no
3. When something isn't impossible but doesn't make sense? _____
4. Name 2 words for something that should happen. _____
5. Name 2 words when it probably won't happen. _____
6. What's a word for something that has to happen? _____

Is it a long shot prediction or an educated guess?

7. **Score exactly 13 points in basketball.**   Long Shot   Educated Guess

8. **Draw an ace out of a deck of cards.**   Long Shot   Educated Guess

9. **Get to be an astronaut.**   Long Shot   Educated Guess

10. **Get to be a fast food worker.**   Long Shot   Educated Guess

11. **Chances you'll go Starbucks this week.**   Long Shot   Educated Guess

12. **Chances you'll go shopping this week.**   Long Shot   Educated Guess

Ch 8 Ls 3 How Probability Uses Fractions   73

_____ #1 #2 ____/ 12 #3 ____/ 4 R ____/ 8 Total ____/ 24 _____
             Name   Checker

#1 1. Name 2 things needed for a math event. _____

2. What does the denominator show about an event's chances? _____

3. What does the 1 show you? _____

4. What makes a fair spinner? _____

5. What fraction shows the odds of flipping a coin and getting tails? _____

6. What does it equal when you add all the chances for an event? _____

#2 1. Make a fraction for these events happening.   If you flip a coin 1 time, what is the fraction for tails happening?

2. What are the fractions for both heads and tails. What does it equal?   ⬜ chances tails / total chances

___ chances heads ___ chances tails   — + — = —

3. What fraction is this?   If you toss a 6 sided die, what is the fraction for 5 happening?

4. Find a fraction for the 1, 2, 3, 4, and 6 happening?   ⬜ chances for 5 / total chances

___ chances heads ___ chances tails   — + — = —

5. What fraction makes this happen out of a deck of cards?   If you choose a card out of 52, what is the fraction that it happens?

6. What are the fractions for the other cards. What does it equal?   ⬜ chances aces / total chances

___ chances heads ___ chances tails   — + — = —

#3 1. Make a spinner for 3 yellow blank, 4 purple blanks, and 1 white blank.

Make a fraction chance and a word for each event.

What are the fractions for each? Calculator?   yes   no

___ + ___ + ___ = ___

2. What is the range of numbers of colors? _____

3. Make a spinner with 3 colors that has 5 red blanks, 3 white blanks, and the rest are blue.

What are the fractions for each?

___ + ___ + ___ = ___

4. Make a spinner with 3 colors that has 1 red blank, 3 white blanks, and the rest are blue.

What are the fractions for each?

___ + ___ + ___ = ___

Review 1. Name 2 things needed for a math event. _____   Calculator?   yes   no

2. What does the denominator show about an event's chances? _____

3. What does the 1 show you? _____

4. What makes a fair spinner? _____

5. What fraction shows the odds of flipping a coin and getting tails? _____

6. What does it equal when you add all the chances for an event? _____

7. Make a spinner with 3 colors that has 5 red, 2 white, and the rest blue.

What are the fractions for each?

___ + ___ + ___ = ___

8. Make a spinner with 3 colors that has 5 blue, 3 white, and the rest red.

What are the fractions for each?

___ + ___ + ___ = ___

Ch 8 Ls 2 How Probability Uses Percent   75

_____ #1 #2 ____/ 14   #3 #4 ____/ 8   R ___/ 7   T ____/ 29 _____
Name   Checker

#1 1. What percents does prediction use? _____
   2. What is the "other" percent? _____
   3. What percent tells about an event that can't possibly happen? _____
   4. What percent tells about an event that has to happen? _____

   Use the percent line to tell about each word.
   5. Which percents go with this word? Draw them in.   **Maybe**

   0%   50%   100%

   6. Which percents go with this word?   **Should happen**

   0%   50%   100%

   7. Which percents go with this word? Draw it in.   **Certain**

   0%   50%   100%

#2 1. 5 red, 2 blue, 3 yellow What is the denominator for each fraction? _____
   2. 5 red, 2 blue, 3 yellow How do you find the percents? _____
   3. Take a red. What 3 things changed? _____
   4. Make an equation.   **A bag of candy has 4 green, 6 red, and 10 blue.
                           What are the chances of picking the colors?**

   5. Change to percents.

   6. Jeff took all the blue ones.
      What are the fractions now?

   7. Change to percents.

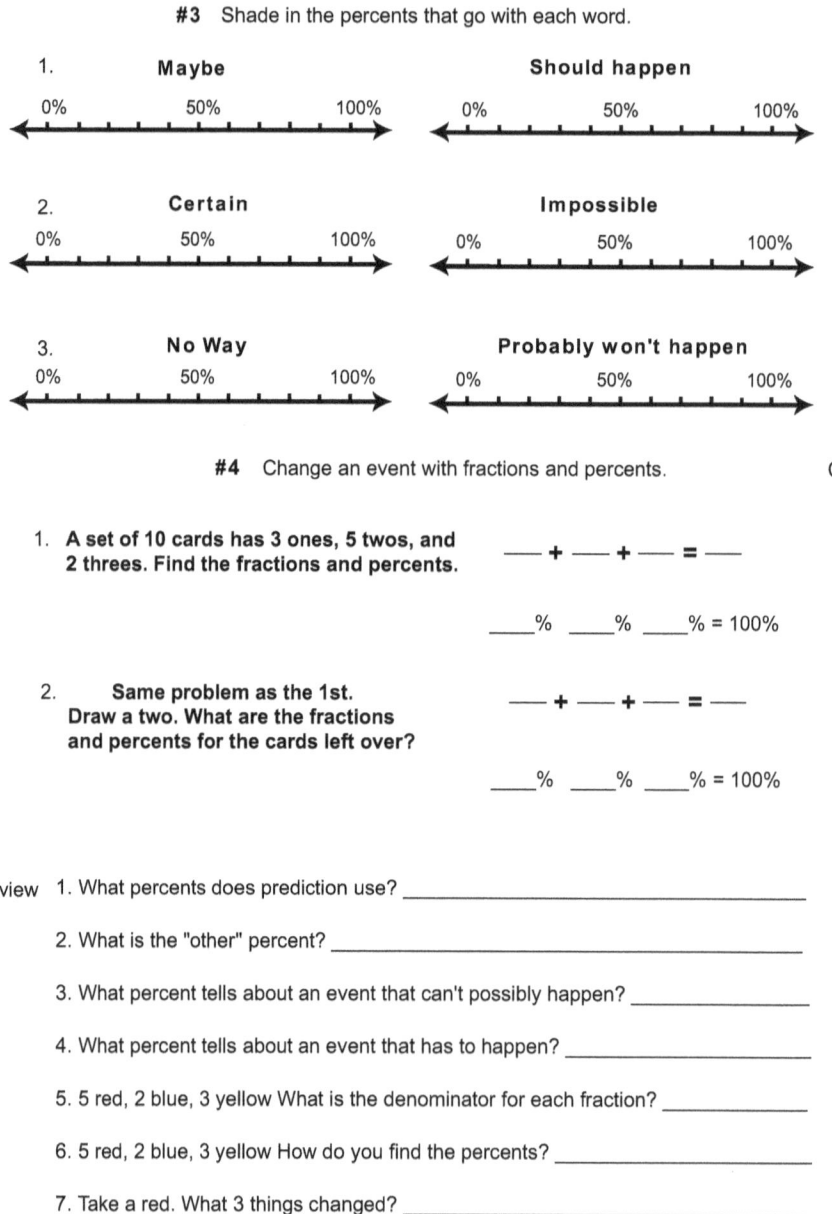

#3 Shade in the percents that go with each word. Calculator? yes no

1. Maybe
0% 50% 100%

Should happen
0% 50% 100%

2. Certain
0% 50% 100%

Impossible
0% 50% 100%

3. No Way
0% 50% 100%

Probably won't happen
0% 50% 100%

#4 Change an event with fractions and percents. Calculator? yes no

1. A set of 10 cards has 3 ones, 5 twos, and 2 threes. Find the fractions and percents.

___ + ___ + ___ = ___

___% ___% ___% = 100%

2. Same problem as the 1st. Draw a two. What are the fractions and percents for the cards left over?

___ + ___ + ___ = ___

___% ___% ___% = 100%

Review 1. What percents does prediction use? _____

2. What is the "other" percent? _____

3. What percent tells about an event that can't possibly happen? _____

4. What percent tells about an event that has to happen? _____

5. 5 red, 2 blue, 3 yellow What is the denominator for each fraction? _____

6. 5 red, 2 blue, 3 yellow How do you find the percents? _____

7. Take a red. What 3 things changed? _____

Review Problems

_____ #1 #2 ____ / 16 #3 ____ / 12 Total ____ / 28
              Name

#1 1. Probability _____

2. Prediction _____

3. Impossible _____

4. No Way _____

5. Possibly _____

6. Certain Event _____

7. Unfair Spinner _____

8. Event _____

Calculator?
yes   no

#2
Story Problems

1. A bag contains 3 red, 8 yellow, and 5 purple marbles. What is the probability of pulling out a red marble? _____

2. Make a spinner with 3 colors that has 5 red, 2 white, and 3 blue. What are the chances of flipping a coin heads and turning white. _____

3. The sides of a number cube have the numbers 1, 3, 6, 9, 12, and 15. If the cube is thrown once, what is the probability of rolling a 3? _____

4. There are 7 items to put on your sandwich. You can only put 3 of them on. How many combinations of items can you make? _____

5. A set of 12 cards has 3 ones, 5 twos, and 4 threes. Find the fractions and percents.
   Fraction _____
   Percent   _____

6. A snow storm dumped 23 cm of snow in a 10 hr period. How many inches were falling per hour? _____

7. TJ has 3 oranges and 2 apples in a bag. He's waiting for a bus and, without looking, picks 2 fruit and eats them. What are the chances it's 2 apples? _____

8. A Prior Car drove 180 kilometers on 8 liters of gas. How far should it be able to go on a full, 18 liter, tank? _____

#3 Story Problems

1. Mitul's teacher had 6 calculators for 24 students to use. If the students are chosen at random. What are the chances that Jake will get one? _____ Calculator? yes no

2. The rental car company had 12 cars and 6 minivans available to rent. If the next customer picks a vehicle at random, what are the chances a car is chosen? _____

3. A magician does a magic trick where he picked one card from a 52 card deck. What's the chance a black card will be chosen? _____

4. A 6 sided die is numbered 1 to 6. If the cube is thrown once, what is the probability of rolling an odd number in percent? _____

5. Mr K is getting up earlier in the morning. He wakes up at 9, but has to get up at 7:50. If he wakes up 10 min earlier each morning, how long will it take? _____

6. A bag contains 10 blue, 20 white, and 10 grey marbles. You pick one without looking. What is the probability it is white or blue? _____

7. If 2 six sided die are rolled, what is the probability that the number rolled will be double 5s? _____

8. Zara bought a new bike for which the probability of having a defective sprocket is 0.2%. What is the probability of not having a defective wheel? _____

9. You ask a friend to think of a number from 4 to 14. What is the probability that his number will be 7 in percent? _____

10. TJ is making a salad from 2 types of fruit. He has melons, strawberries, grapes, oranges, and bananas. How many combinations could he make? _____

11. A bag contains 6 green, 8 blue, and 11 red shirts. You ask a friend to pick one without looking. What is the probability that the shirt will be green? _____

12. 12% of Indians who own autos, own trucks. If you ask a random person whether he or she owns a truck, what is the probability that the person does not own a truck? (percent answer) _____

Ch 9 Ls 1 Probability With More Chances   79

_____ #1 #2 ____/11 #3 #4 ____/12 R ___/9 Total ____/30 _____
Name   Checker

#1 1. What are multiple events? _____
    2. How do you solve multiple events? _____
    3. What do you use to find the possibilities with 3 coins? _____
    4. Is 1 coin the base or exponent? _____
    5. How do you find percent? _____

    6. What exponent finds this chance?   **If you flip 3 coins, what are the chances of all heads?**

    What percents show the chances? $\dfrac{\quad}{\quad}$

    Divide _____ .   ____% chance of all heads.

    7. What exponent finds this chance?   **If you flip 4 coins, what are the chances of 2 heads 2 tails?**

    What percents show the chances? $\dfrac{\quad}{\quad}$

    Divide _____ .   ____% chance of all heads.

#2 1. What is the name for this rule? _____
    2. How do you find probability with different kinds of events? _____
    3. What are independent events? _____ .
    4. Make a fraction for an event,   **5 shirts and 3 pants hang equally in a closet.**
       then a percent for the events.   **What are the chances of picking 1 of each of them?**

$\dfrac{\quad}{\quad} \times \dfrac{\quad}{\quad} = \dfrac{\quad}{\quad}$ Change the fraction to a percent.

$\dfrac{\quad}{\quad} = $ ____%

#3 Use sets of cards to show what happens with the exponent.   Calculator? yes no

1. Make fractions for this event.   If you flip a die 2 times, what is the fraction for double 6s happening?

   $\frac{\quad}{\quad}$ is $\frac{\quad}{\quad} + \frac{\quad}{\quad} = \frac{\quad}{\quad}$ What percents? ____% + ____% = ____%

2. Make fractions for this event.   If you choose a card from 52, what is the fraction for kings happening?

   $\frac{\quad}{\quad} + \frac{\quad}{\quad} = \frac{\quad}{\quad}$ What percents? ____% + ____% = ____%

3. Make fractions for this event.   If you flip a coin 5 times, what is the fraction for 5 heads happening?

   $\frac{\quad}{\quad}$ is $\frac{\quad}{\quad} + \frac{\quad}{\quad} = \frac{\quad}{\quad}$ What percents? ____% + ____% = ____%

Review   1. What are multiple events? _____   Calculator? yes no

2. How do you solve multiple events? _____

3. What do you use to find the possibilities with 3 coins? _____

4. Is 1 coin the base or exponent? _____

5. How do you find percent? _____

6. What is the name for this rule? _____

7. How do you find probability with different kinds of events? _____

8. What are independent events? _____

9. Make fractions for this event.   If you choose a card from 52, what is the fraction for not jacks happening?*

   What percents show the chances? $\frac{\quad}{\quad} + \frac{\quad}{\quad} = \frac{\quad}{\quad}$

   ____% + ____% = ____%

   *Use rounding.

Ch 9 Ls 3 Dependent Events   81

_____ #1 #2 ____/10 #3 ____/5 R ___/10 T ____/25 _____
         Name   Checker

#1 1. What are independent events? _____
        2. What are dependent events? _____
        3. How do you solve dependent events? _____
        4. What numbers make a factorial? _____
        5. Pick 1 card out of 5. How does a factorial solve it? _____

#2 1. What 2 facts multiply to get this factorial?   **5!**

        5 x 4 = _____   3 x 2 x 1 = _____   Multiply them is _____.

        2. What factorial can you use to get this one?   **6!**

            5! = _____, so multiply _____ x _____ = _____

        Decide if it uses independent or dependent events.

        3. How do you solve it?   Pick a card from 10 of them and put it back, then
                                   pick again. What are the chances of picking it twice?

        Circle one **Dependent**   **Independent**   _____ x _____ = _____

        4. How do you   Toss a die of 6 and get a number you chose. What are the
           solve it?   chances of tossing another one and getting the same number?

        Circle one **Dependent**   **Independent**   _____ x _____ = _____

        5. How do you   Spin a spinner with 6 chances and flip a coin. What are
           solve it?   the chances of getting 1 of them and flipping heads?

        Circle one **Dependent**   **Independent**   _____ x _____ = _____

#3 Decide if it uses independent or dependent events.

Calculator? yes no

What do you multiply?

1. A boy has 5 red cubes and 5 blue cubes. He chooses 1 and then puts it back. If he chooses twice, what are the chances they are both red?

____ x ____ = ____

2. A boy has 5 red cubes and 5 blue cubes. He chooses 1 and then keeps it. If he chooses twice, what are the chances they are both red?

____ x ____ = ____

3. TJ has 3 oranges and 2 apples in a bag. He's waiting for a bus and, without looking, picks 2 and eats them. What are the chances it's 2 apples?

____ x ____ = ____

4. Same problem. What are the chances it's 2 oranges?   ____ x ____ = ____

5. Same problem. What are the chances both fruits are the same kind?

____ x ____ = ____

Review 1. What are independent events? _____

Calculator? yes no

2. What are dependent events? _____

3. How do you solve dependent events? _____

4. What numbers make a factorial? _____

5. Pick 1 card out of 5. How does a factorial solve it? _____

What are these factorials?

6. **4!** _____

7. **5!** _____

8. **7!** _____

9. **8!** _____

10. **10!** _____

Ch 9 Ls 4 Permutations 83

_____ #1 #2 ____/ 10 #3 ____/ 6 R ____/ 8 Total ____/ 24 _____
       Name   Checker

#1 1. What is a permutation? _____
   2. What kind of number solves a permutation? _____
   3. How do you write the short version for a permutation? _____
   4. What does **P (4, 2)** show? _____

   #2 1. What does it multiply? Solve it.   **P(7, 2)**

                                          ____ x ____ = ____

   2. What is this permutation?   **P(5, 3)**

                                     ____ x ____ x ____ = ____

   3. How does 1 more number change it?   **P(6, 4)**

                                     ____ x ____ x ____ = ____

   4. What does it multiply? Solve it.   **P(9, 2)**

                                          ____ x ____ = ____

   5. What do you multiply?   There are 9 horses. Top 2 receive a prize.
                               What are the chances of getting a prize?

                                      _____

   6. What do you multiply?   An artist drew 8 pictures, which he'll give to 3 of
                               his patrons. How many ways can they be given?

                                      _____

#3 What does it multiply? Solve it.   Calculator? yes no

1.   P(4, 2)   P(7, 3)

   _____ = ___   _____ = ___

2.   P(8, 2)   P(5, 3)

   _____ = ___   _____ = ___

3.   P(9, 2)   P(6, 3)

   _____ = ___   _____ = ___

Review
1. What is a permutation? _____   Calculator?
2. What kind of number solves a permutation? _____   yes no
3. How do you write the short version for a permutation? _____
4. What does P (4, 2) show? _____

5. Find these permutations.   P(11, 2)   P(10, 3)

   ___ x ___ = ___   ___ x ___ x ___ = ___

6. Find these permutations.   P(7, 2)   P(8, 3)

   ___ x ___ = ___   ___ x ___ x ___ = ___

7. What do you multiply?   Ms K has 4 pieces of candies. There are 8 students could get them. Chances of getting 1?

   _____

8. What do you multiply?   There are 10 students names in a jar. 4 of them get a treat. How many different combinations of names are drawn?

   _____

Ch 9 Ls 5 Combinations 85

_____ #1 #2 ____/10 #3 ____/5 R ___/8 T___/23 _____
Name   Checker

#1 1. What is a combination? _____

2. What 2 steps solve a combination? _____

3. How do you solve the combination **C (4, 2)**? _____

4. Why is **what's taken** important? _____

#2 1. What is the 1st step?   C(5, 2)

____ x ____ = ____   What is the 2nd step?

____ ÷ ____ = ____

2. All 1 step. What does it multiply and divide?   C(6, 2)

____ x ____ = ____   ____ ÷ ____ = ____

3. Multiply and divide it.   C(7, 2)

____ x ____ = ____   ____ ÷ ____ = ____

4. Multiply and divide it.   C(5, 3)

____ x ____ = ____   ____ ÷ ____ = ____

5. What do you multiply?   There are 10 different pizza toppings at the pizza shop. How many different can 2 toppings be chosen, none twice.

____ x ____ = ____   ____ ÷ ____ = ____

6. What do you multiply?   Mr K picked out 7 books from the libary, but they only let him take 2 out. How many possible selections can he choose?

____ x ____ = ____   ____ ÷ ____ = ____

#3 Solve these combinations.   Calculator? yes no

1. C(4, 2) ___ x ___ = ___   ___ ÷ ___ = ___
2. C(5, 2) ___ x ___ = ___   ___ ÷ ___ = ___
3. C(5, 3) ___ x ___ = ___   ___ ÷ ___ = ___
4. C(6, 2) ___ x ___ = ___   ___ ÷ ___ = ___
5. C(6, 3) ___ x ___ = ___   ___ ÷ ___ = ___

Review 1. What is a combination? _____   Calculator?
2. What 2 steps solve a combination? _____   yes no
3. How do you solve the combination C (4, 2)? _____
4. Why is **what's taken** important? _____

5. There are 7 items to put on your sandwich. You can only put 3 of them on. How many combinations of items can you make?
    ___ x ___ = ___
    ___ ÷ ___ = ___

6. There are 5 balls. You can pick 3 of them. How many different ways can be chosen?
    ___ x ___ = ___
    ___ ÷ ___ = ___

7. To win the lottery, you must correctly select 3 numbers from 20 numbers, the order doesn't does not matter. How many different selections are possible?
    ___ x ___ = ___
    ___ ÷ ___ = ___

8. A 3 person committee is to be elected from an organization's membership of 12 people. How many different committees are possible?
    ___ x ___ = ___
    ___ ÷ ___ = ___

Review Problems 87

_____ #1 #2 ____/15 #3 ____/11 Total ____/27
                Name

#1
1. Exponent Problem _____
2. Fundamental Counting Principle _____
3. Dependent Event _____
4. Factorial _____
5. Permutation _____
6. Combination _____

Calculator?
yes   no

#2
Story Problems

1. Suppose that 6 people enter a swim meet. How many ways could the gold, silver, and bronze medals be awarded? (no ties)   _____

2. Twenty people bought raffle tickets to enter a random drawing for 2 prizes. How many ways can 2 names be drawn for 1st and 2nd prize?   _____

3. If you roll a die 2 times, what is the fraction of 2 fours happening?   _____

4. A coach must choose how to line up his three starters from a team of 10 players. How many different ways can the he choose the starters?   _____

5. Mitul is making a salad from five different fruits: bananas, oranges, apples, pineapples, and lemons. If he uses two fruits, how many salads can he make?   _____

6. How many different 2 letter passwords can be created for a software access if no letter can be used more than once?   _____

7. How many different ways you can elect a president and vice president of a committee if you have 11 people to choose from.   _____

8. If you flip a coin 3 times, what is the fraction of 3 heads happening?   _____

9. A disc jockey has to choose three songs for a show. If there are 10 songs that are long enough, then how many ways can he choose and arrange to play them?   _____

#3 Story Problems   Calculator?
                                                       yes   no

1. A rental car company had 12 cars and 6 minivans
   available to rent. If the next customer picks a vehicle   _____
   at random, what are the chances a car is chosen?

2. You and 2 friends go to a restaurant for soup. The
   menu has 6 types of soups. What is the probability   _____
   that each of you orders a different type?

3. Mr K is getting up earlier in the morning. He wakes
   up at 9, but has to get up at 7:50. If he wakes up   _____
   10 min earlier each morning, how long will it take?

4. To win a lottery, I must correctly select 2 numbers
   from 30 numbers. How many different selections   _____
   are possible?

5. When a six sided die is rolled 2 times, what is the
   probability that the number rolled will be 2/2?   _____

6. I roll a six-sided die and toss a coin. What is the
   probability I get a tails and a 3?   _____

7. I draw a card from a poker deck, replace it, and
   draw another. What is the probability I get 2 hearts?   _____

8. The running club with 7 members is set to choose
   2 leaders, president and co-president How many   _____
   ways can those offices be filled?

9. You are asked to list the 3 best movies you have
   seen this year. You saw 10 movies during the year.   _____
   How many ways can they be ranked?

10. A 2-person committee is to be elected from an
    organization's membership of 40 people. How   _____
    many different committees are possible?

11. You draw a card from 25 cards numbered from
    1 to 25. What is the probability of drawing a   _____
    squared number? (1, 4, 9, 16, 25)

Ch 10 Ls 1   Different Kinds of lines and Angles   89

_____ #1 #2 ____/10 #3 ____/12 R ___/ 9 Total ____/31 _____
          Name   Checker

#1 1. There are 3 kinds of lines. What makes them different? _____
2. What are acute angles? _____ Obtuse angles? _____
3. How many degrees are in a right angle? _____
4. How many degrees are in 1 hour of a clock? _____

#2 1. What kind of line is it?

2. Is it positive or negative?
How many 90s?

positive   negative   It's a _____ angle.

3. Is it positive or negative?
How many 90s?

positive   negative   It's a _____ angle.

4. Is it positive or negative?
How many 90s?

positive   negative   It's a _____ angle.

5. Is it positive or negative?
How many 90s?

positive   negative   It's a _____ angle.

6. What kind of line is it?

#3 What does each picture show?   Calculator? yes   no

1. _____ degrees   _____

2. _____   _____

3. _____   _____ degrees

4. _____ degrees   _____

5. _____ degrees   _____

6. _____ kind of angle   _____ degrees

Review 1. There are 3 kinds of lines. What makes them different? _____   Calculator? yes   no

2. How are acute and obtuse angles different? _____

3. What are these lines called?

_____   _____

4. How many degrees are in a right angle? _____

5. How many degrees are in half way around a clock? _____

6. What names are these lines?

_____   _____   _____

Ch 10 Ls 2 Counting Angles Like a Clock   91

_____ 1# #2 ____/ 9   #3 ____/ 10   R ____/ 10   Total ____/ 29 _____
           Name   Checker

#1 1. 1 minute on a clock is how many degrees? _____

2. One hour on a clock is how many degrees? _____

3. Name 2 steps to count this angle. _____

4. What is half of a 90 degree angle? _____

#2 1. How many degrees are 1 and 4 on these clocks?

_____ deg   _____ deg

2. How many degrees are 1:30 and 2:30?

_____ deg   _____ deg

3. How many degrees are 10 and 11 on these clocks?

_____ deg   _____ deg

4. Find the degrees in 10:30 and 11:30.

_____ deg   _____ deg

5. Find the degrees in 7:00 and 7:30.

_____ deg   _____ deg

#3 Decide the 30 degree angles.   Calculator?
                                                                yes   no

1. _____ degrees   _____ degrees

2. _____ degrees   _____ degrees

3. _____ degrees   _____ degrees

4. _____ degrees   _____ degrees

5. _____ degrees   _____ degrees

Review 1. 1 minute on a clock is how many degrees? _____   Calculator?
        2. One hour on a clock is how many degrees? _____   yes   no
        3. Name 2 steps to count this angle. _____
        4. What is half of a 90 degree angle? _____

Decide these 45 degree angles.

5. _____ degrees   _____ degrees

6. _____ degrees   _____ degrees

7. _____ degrees   _____ degrees

Ch 10 Ls 3 How a Compass Uses Angles 93

_____ #1 #2 ____/ 15 #3____/ 8 R ___/ 7 Total ____/ 30 _____
Name   Checker

#1 1. What are the 4 basic directions? _____
   2. Which direction is between north and east? _____
   3. What are the degrees for northeast? ____ degrees   Northwest? ____ degrees
   4. What are the degrees for southeast? ____ degrees   Southwest? ____ degrees
   5. What do you use to find a side of a right triangle? _____

   #2 1. What direction is it?   Go south 20 ft and west 10 ft.

      2. How long is C?
         What's the equation?   _____

         Solve the next step. Square the numbers.   ___^{2} + ___^{2} = ___^{2}

      3. What's the answer?   ___ + ___ = ___

                                       _____

      4. What direction is it?   Go north 10 ft and east 60 ft.

      5. How long is C?
         What's the equation?   _____

         Solve the next step. Square the numbers.   ___^{2} + ___^{2} = ___^{2}

      6. What's the answer?   ___ + ___ = ___

                                       _____

#3 What direction does each arrow point?   Calculator?
yes   no

1. ↖ _____   → _____

2. ↑ _____   ↘ _____

3. ← _____   ↗ _____

4. ↙ _____   ↓ _____

Review 1. What are the 4 basic directions? _____   Calculator?
yes   no
2. Which direction is between north and east? _____
3. What are the degrees for northeast? ____ degrees   Northwest? ____ degrees
4. What are the degrees for southeast? ____ degrees   Southwest? ____ degrees
5. What do you use to find a side of a right triangle? _____

What direction is it? Find distance back to 0.

6. **Go south 20 ft and west 10 ft.**

_____
Direction

$\underline{}^2 + \underline{}^2 = \underline{}^2$

$\underline{} + \underline{} = \underline{}$

_____

7. **Go north 30 ft and east 30 ft.**

_____
Direction

$\underline{}^2 + \underline{}^2 = \underline{}^2$

$\underline{} + \underline{} = \underline{}$

_____

Ch 10 Ls 4 Complementary and Supplementary Angles   95

_____ #1 #2 ____/14 #3 #4 ____/12 R ___/ 9 Total ____/30 _____
                Name   Checker

#1 1. What are complementary angles? _____

2. What are supplementary angles? _____

3. What makes complementary angles? Find it.   **50 degrees**

They add to ____ degrees.   50 + ___ = 90

4. What makes supplementary angles?   **70 degrees**

They add to ____ degrees.   70 + ___ = 180

#2 1. What are opposite angles? _____

2. Name the degrees in a right angle. _____

3. What angles add to get 90 and 180?   30 + 2x = 90   30 + 2x = 180

2x = ___   2x = ___

x = ___   x = ___

4. What are both angles?   b⟨a / 40 degrees / c⟩

Angle a uses supplemental angle and they both use opposite angles.

40 degrees / 140° / 140° / 40 degrees

5. What is this angle?

x° / 30° / ⌐

30 + x = 90
x = 60

#3 Find the Complementary angles.   Calculator? yes no

1. 50 degrees: _____   70 degrees: _____   -10 degrees: _____

2. 20 degrees: _____   45 degrees: _____   -40 degrees: _____

3. 8 degrees: _____   18 degrees: _____   -70 degrees: _____

#4 Supplementary angles.

1. 30 degrees: _____   120 degrees: _____   -20 degrees: _____

2. 45 degrees: _____   70 degrees: _____   30 degrees: _____

3. 90 degrees: _____   150 degrees: _____   -35 degrees: _____

Review 1. What are complementary angles? _____   Calculator? yes no

2. What are supplementary angles? _____

3. What are opposite angles? _____

4. How many degrees are in right angles? _____

Solve for the complementary or supplementary angles for these angles.

5. $10 + 4x = 90$ $30 + 5x = 180$ $20 + 7x = 90$

6.
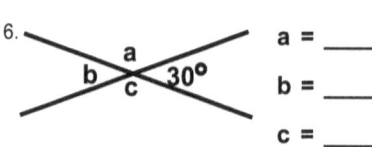
a = _____°
b = _____°
c = _____°

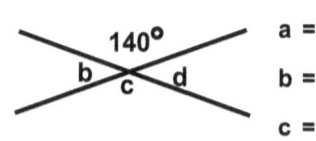
a = _____°
b = _____°
c = _____°

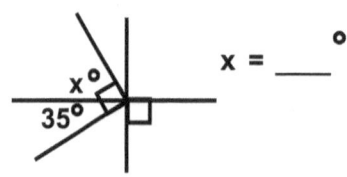
x = _____°

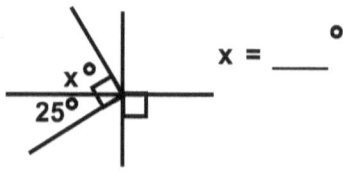
x = _____°

Review Problems 97

_____ #1 #2 ____ /22 #3 to #6 ____ / 22 Total ____ / 44
              Name

#1 1. Line _____
    2. Line Segment _____
    3. Ray _____
    4. Angle _____
    5. Right Angle _____
    6. Compass _____
    7. Complementary Angle _____
    8. Supplementary Angle _____

#2 What does each picture show?   Calculator?
                                                     yes   no

1.   _____   _____ degrees

2.   _____ degrees   _____ degrees

3.   _____   _____

4.   _____   _____ degrees

5.   _____   _____
     kind of angle   kind of angle

6.   _____ degrees   _____

7.   _____   _____

#3 What direction is it? Find distance back to 0.   Calculator? yes no

1. **Go south 40 ft and west 10 ft.**

   _____
   Direction

   $\underline{}^2 + \underline{}^2 = \underline{}^2$

   $\underline{} + \underline{} = \underline{}$

   _____

2. **Go north 50 ft and east 30 ft.**

   _____
   Direction

   $\underline{}^2 + \underline{}^2 = \underline{}^2$

   $\underline{} + \underline{} = \underline{}$

   _____

#4 Find the Complementary angles.   Calculator? yes no

1. 50 degrees: ____   70 degrees: ____   10 degrees: ____
2. 20 degrees: ____   45 degrees: ____   40 degrees: ____

#5 Supplementary angles.

1. 30 degrees: ____   120 degrees: ____   20 degrees: ____
2. 45 degrees: ____   70 degrees: ____   30 degrees: ____

#6 How long is each angle? (They're 30 degrees angles.)   Calculator? yes no

1. a _____ degrees   a _____ degrees

2. a _____ degrees   a _____ degrees

3. a _____ degrees   a _____ degrees

4. a _____ degrees   a _____ degrees

Ch 11 Ls 1 Triangles and Quadrilaterals   99

_____   #1 #2 ____/ 16   #3 ____/ 8   R ___/ 16   T ____/30 _____
                Name   Checker

#1 1. How many degrees does a triangle's angles add to? _____

2. What's the name for a triangle with 3 equal angles? _____

3. What makes a right triangle? _____

4. What makes an obtuse triangle? _____

5. What kinds of triangles are these?

_____   _____

6. What kinds of triangles are these?

_____   _____

#2 1. How many degrees are in any quadrilateral? _____

2. How is a rhombus different from a square? _____

3. How is a parallelogram different from a rectangle? _____

4. How is a trapezoid different from a parallelogram? _____

5. What kinds of quadrilaterals are these?

_____   _____

6. What kinds of quadrilaterals are these?

_____   _____

#3 Quadrilaterals   Calculator?
                                                                                    yes   no

1. How many degrees are in any quadrilateral? _____
2. How is a rhombus different from a square? _____
3. How is a parallelogram different from a rectangle? _____
4. How is a trapezoid different from a parallelogram? _____

What's the name for each shape?

5. _____   6. _____

7. _____   8. _____

Review 1. How many degrees does a triangle's angles add to? _____
2. What's the name for a triangle with 3 equal angles? _____
3. What makes a right triangle? _____
4. What makes an obtuse triangle? _____

5. _____   6. _____

7. _____   8. _____

Decide what shape goes with each set of angles. (Can be more than 1)

9.   30 60 90 _____   90 90 90 90 _____

10. 50 50 80 _____   80 100 80 100 _____

11. 42 64 74 _____   90 70 90 110 _____

12. 60 60 60 _____   20 60 100 _____

Ch 11 Ls 2 Shapes with More Sides 101

_____ #1 #2 ____/ 17 #3 ____/ 8 R ___/ 10 T ____/ 35 _____
           Name   Checker

#1 1. What shape has 6 equal sides? _____

2. Name 1 thing in the world that has 6 equal sides. _____

3. What shape has 8 equal sides? _____

4. Name 1 thing in the world that has 8 equal sides. _____

5. What shape has 10 equal sides? _____

6. Name the shapes.

_____   _____

7. What shapes are they?

_____   _____

#2 1. What shape has 9 equal sides? _____

2. What building has 5 sides and where is it located? _____

3. What shape has 7 equal sides? _____

4. How are odd shapes different from even shapes? _____

5. Name the shapes.

_____   _____

6. What shapes are they?

_____   _____

#3 Name the shapes with equal sides.   Calculator?
                                          yes   no

1. _____ _____

2. _____ _____

3. _____ _____

4. _____ _____

Review 1. What shape has 6 equal sides? _____

2. Name 1 thing in the world that has 6 equal sides. _____

3. What shape has 8 equal sides? _____

4. Name 1 thing in the world that has 8 equal sides. _____

5. What shape has 10 equal sides? _____

6. What shape has 9 equal sides? _____

7. What building has 5 sides and where is it located? _____

8. What shape has 7 equal sides? _____

9. What shape has 3 and 4 sides? _____

10. How are odd shapes different from even shapes? _____

Ch 11 Ls 3 Symmetry and Congruency 103

_____ #1 #2 ____ / 10 #3 ____ / 20 R ____ / 11 T ____ / 41 _____
Name   Checker

#1   1. What is symmetry? _____

   2. Name 3 kinds of symmetry. _____

   3. What kind of symmetry does a B have? Draw it in.   **B**

B _____

   4. Name the symmetry for W and draw it.   **W**

W _____

   5. Name the symmetry for X and draw it.   **X**

X _____

#2   Simulated Polygons

1. What is a Similar Polygon? _____

2. Are these polygons similar or not?

   Circle one.   **Yes   or   No**

3. Are these polygons similar or not?

   Circle one.   **Yes   or   No**

#3 Which kind of symmetry do you use?   Calculator?
                                                          yes   no

1. **B**   **W**   **I**   ◯   **M**

2. ◻   **H**   **V**   **X**   **A**

Draw the
symmetry.
3. ♡   **T**   **K**   🦋   **O**

4. **E**   **C**   **D**   **J**   ☃

Review 1. What is symmetry? _____

2. Name 3 kinds of symmetry. _____

3. What does it mean to be congruent? _____

4. What's different between congruent and symmetry? _____

5. What is a Similar Polygon? _____

Are these polygons similar or not?

6.   ◯ ⬭   Yes or No   ◻ ◻   Yes or No

7.   △ △   Yes or No   ✦ ✦   Yes or No

8.   ☆ ☆   Yes or No   ▯ ▯   Yes or No

Ch 11 Ls 4 Translation Changes Things   105

_____ #1 #2 ____/ 11 #3 ____/ 6 R ___/ 7 Total ____/ 24 _____
　　　　Name　　　　　　　　　　　　　　　　　　　　　　　　　　　　　　　Checker

#1 1. What does it mean to be congruent? _____
   2. What's different between congruent and symmetry? _____
   3. If a shape is turned from the other, is it still congruent? _____

   4. Which of these are congruent and explain why?   #1.   #2.   #3.

#2 1. What is the name for an object that moves 1 direction? _____
   2. What is the name for how a ketchup bottle moves? _____
   3. How does a mirror show the 3rd way an object moves? _____
   4. How does a wheel spinning around show the 4th way? _____

   5. What kind of movement is this?

---

   6. What kind of movement is this?

---

   7. What kind of movement is this?

#3 Solve the story problems.   Calculator?
                                            yes   no

1. On a map with a scale of 1 cm to 2 kilometers, _____
   the distance between two citiess is 5 centimeters.
   What is the real distance between the two cities? _____

2. On an architect's blueprint, the front of a bridge _____
   measures 30 cm. The scale of the blueprint is
   1 cm to 2 meters. How long will the front of the _____
   real bridge be?

3. The model of a jet has a wingspan of 18 cm.
   The model has a scale of 1 cm to 3 meters. _____
   What is the wingspan of the actual jet?
                                              _____

4. The drawing for a building has a scale of 1 cm to
   2 meters. The building in the drawing has a height _____
   of 20 cm. How tall will the real building be?
                                              _____

5. A model of a rocket has a scale of 1 dm to
   4 meters. If the model rocket is 30 decimeters _____
   tall, how tall was the rocket?
                                              _____

6. A model of a 6-cylinder gasoline engine is built
   on a scale of 1 cm is 4 decimeters. If the length _____
   of the model engine is 8 centimeters, how long
   is the engine?   _____

Review 1. What does it mean to be congruent? _____

        2. What's different between congruent and symmetry? _____

        3. If a shape is turned from the other, is it still congruent? _____

        4. What is the name for an object that moves 1 direction? _____

        5. What is the name for how a ketchup bottle moves? _____

        6. How does a mirror show the 3rd way an object moves? _____

        7. How does a wheel spinning around show the 4th way? _____

Vocabulary 107

_____ #1 #2 ____ / 21 #3 #4 #5 ____ / 20 Total ____ / 41
              Name

#1 1. Equilateral Triangle _____
    2. Right Triangle _____
    3. Isoceles Triangle _____
    4. Acute Triangle _____
    5. Obtuse Triangle _____
    6. Scalene Triangle _____
    7. Square _____
    8. Rhombus _____
    9. Parallelogram _____
    10. Trapezoid _____
    11. Hexagon _____
    12. Octagon _____
    13. Pentagram _____
    14. Symmetry _____
    15. Congruent _____
    16. Translation _____

Calculator? yes no

#2 Name the shapes.

1. _____   2. _____   3. _____
4. _____   5. _____   6. _____
7. _____   8. _____   9. _____
10. _____ 11. _____ 12. _____

#3 Which kind of symmetry do you use?   Calculator?
yes   no

Vertical, horizontal, or diagonal.

1. E   2. C   3. (butterfly)
___   ___   ___

4. D   5. O   6. X
___   ___   ___

7. □   8. M   9. ♡
___   ___   ___

#4 Are these polygons similar or not?   Calculator?
yes   no

1. (two ovals) Yes No   (two squares) Yes No   (two rectangles) Yes No

2. (two triangles) Yes No   (two 4-point stars) Yes No   (two 5-point stars) Yes No

#5 Story problems.   Calculator? yes   no

1. On a map with a scale of 1 cm to 2 kilometers, the distance between two citiess is 8 cm. What is the real distance between the two cities?
   _____
   _____

2. On an architect's blueprint, the front of a bridge measures 25 centimeters. The scale of the blue print is 1 cm to 5 meters. How long will the front of the real bridge be?
   _____
   _____

3. The model of a jet has a wingspan of 12 dm. The model has a scale of 1 dm is 1 meter. What is the wingspan of the actual jet?
   _____
   _____

4. The drawing for a building has a scale of 1 cm is 3 meters. The building in the drawing has a height of 20 cm. How tall will the real building be?
   _____
   _____

5. A model of a rocket has a scale of 3 cm to 1 meter. If the model rocket is 18 centimeters tall, how tall was the rocket?
   _____
   _____

Ch 12 Ls 1 Perimeter and Area   109

_____ #1 #2 ____ / 12 #3 ____ / 8 R ___ / 8 T ____ / 28 _____
                Name   Checker

#1 1. What is perimeter? _____

2. Name 2 ways to find perimeter. _____

3. What does area find? _____

4. Name 2 ways to find area. _____

5. How is the answer for area labeled? _____

6. Name 2 ways to find this perimeter.   [6 meters | 4 meters]

Add ___ + ___ + ___ + ___ = ___   Multiply 2 x ___ Add 2 x ___ = ___

7. What is the area for the rectangle?   [10 cm | 4 cm]

Multiply ___ x ___ = ___ square _____

#2 1. What is hidden information? _____

2. What is the hidden information?

   2 m
   [A B]
   7 m
   [8 meters | 4 meters]

3. What is the perimeter and area?   **A. 7 - 4 = ___   B. 8 - 2 = ___**

Perimeter is _____   Area is _____

4. What is the hidden information?

   3 dm
   [A B]
   6 dm
   [12 dm | 3 dm]

5. What is perimeter and area?   **A. 6 - 3 = ___   B. 12 - 3 = ___**

Perimeter is _____   Area is _____

#3 Name 2 ways to find this perimeter. Find area.   Calculator?
                                                                                                   yes   no

1. [rectangle: 4 feet by 3 feet]   Perimeter _____

    Area: _____

2. [rectangle: 6 mm by 2 mm]   Perimeter _____

    Area: _____

3. [rectangle: 7 cm by 6 cm]   Perimeter _____

    Area: _____

4. [rectangle: 7 meters by 3 meters]   Perimeter _____

    Area: _____

Review 1. What is perimeter? _____   Calculator?

2. Name 2 ways to find perimeter. _____   yes   no

3. What does area find? _____

4. Name 2 ways to find area. _____

5. How is the answer for area labeled? _____

6. **Decide 2 different shapes that have 30 cm perimeter.**   ____ + ____ + ____ + ____ = 30 cm

    ____ + ____ + ____ + ____ = 30 cm

7. **Find 2 different shapes that have area 20 square meters.**   ____ x ____ = 20 sq m

    ____ x ____ = 20 sq m

8. **Decide 2 different shapes that have perimeter 40 kilometers.**   ____ + ____ + ____ + ____ = 40 km

    ____ + ____ + ____ + ____ = 40 km

Ch 12 Ls 2 English Square Units 111

_____ #1 #2 ____/10 #3 #4 ____/ 8 R ___/ 8 T ____/ 26 _____
                Name   Checker

#1 1. How many square inches are in a square foot? _____

2. How many square feet are in a square yard? _____

3. How do you change square feet
    to square inches? | 3 feet | 1 foot |

____ square feet   Multiply ____ x ____ = ____ square inches

4. How many square inches in 5 sq ft? | 5 feet | 1 foot |

____ square feet   Multiply ____ x ____ = ____ square inches

5. How do you change to square yards? | 6 feet | 3 feet |

____ x ____ = ____ sq yd   ____ ÷ ____ = ____ sq yd

6. How many square yards is it? | 15 feet | 6 feet |

____ x ____ = ____ sq yd   ____ ÷ ____ = ____ sq yd

#2 1. What sports field is like an acre? _____

2. How many square feet are in an acre? _____

3. Perimeter and area of a room?   **Average room (12 x 13 ft)**

            Perimeter is _____   Area is _____

4. Perimeter and area of a closet?   **A closet (3 x 6 ft)**

            Perimeter is _____   Area is _____

#3 How do you change to square inches?   Calculator? yes no

1. [6 feet × 3 feet]   ___ square feet
   Multiply ___ x ___ = _____ sq inches

2. [8 feet × 4 feet]   ___ square feet
   Multiply ___ x ___ = _____ sq inches

#4 How do you change to square yards?

1. [6 feet × 5 feet]   ___ square feet
   Divide ___ ÷ ___ = _____ sq yards

2. [30 feet × 10 feet]   ___ square feet
   Divide ___ ÷ ___ = _____ sq yards

Review 1. How many square inches are in a square foot? _____   Calculator? yes no

2. How many square feet are in a square yard? _____

3. What sports field is like an acre? _____

4. How many square feet are in an acre? _____

5. **Decide 2 different shapes in feet that have area 100 square feet.** ___ ft x ___ ft   ___ ft x ___ ft

6. **Decide 2 different shapes in feet that have area 90 square yards.** ___ ft x ___ ft   ___ ft x ___ ft

7. **Decide 2 different shapes in feet that have perimeter 24 feet.** (___ ft + ___ ft) x 2
   (___ ft + ___ ft) x 2

8. **Decide 2 different shapes in feet that have perimeter 40 feet.** (___ ft + ___ ft) x 2
   (___ ft + ___ ft) x 2

Ch 12 Ls 3 Metric Square Units 113

_____ #1 #2 ____/ 9   #3 #4 ____/ 8   R ___/ 10   T ____/ 30 _____
   Name   Checker

#1 1. How many square centimeters are in a square meter? _____

2. A square centimeter has how many square millimeters in it? _____

3. How do you change to square centimeters? [4 m] 2 m

Multiply ____ x 10,000 = ____ square cm

4. How many square centimeters? [5 m] 3 m

Multiply ____ x 10,000 = ____ square cm

5. How do you change to square mm? [40 cm] 20 cm

Multiply ____ x 100 = _____ square mm

6. How many square millimeters? [60 cm] 30 cm

Multiply ____ x 100 = _____ square mm

#2 1. How many acres are in a hectare? _____

2. How many meters are in a hectare? _____

3. How do you change 3 hectares to acres?   **3 hectares = ? acres**

What's the answer?   Multiply times 2.5   3 x 2.5 = ____

3 hectares = ___ acres

#3 Change meters to square centimeters.   Calculator? yes no

1. [6 m × 3 m rectangle]   ____ square meters
   Multiply ____ x _____ = _____ sq cm

2. [10 m × 6 m rectangle]   ____ square meters
   Multiply ____ x _____ = _____ sq cm

#4 How do you change to square millimeters?   Calculator? yes no

1. [7 cm × 4 cm rectangle]   ____ square cm
   Multiply _____ x ____ = _____ sq mm

2. [9 cm × 5 cm rectangle]   ____ square cm
   Multiply _____ x ____ = _____ sq mm

Review 1. How many square centimeters are in a square meter? _____   Calculator? yes no
2. A square centimeter has how many square millimeters in it? _____
3. How many acres are in a hectare? _____
4. How does meters make a hectare? _____

Change hectares to acres or estimate acres and football fields.

5. Change 4 hectares to acres.   4 hectares = ____ acres
6. Change 6 hectares to acres.   6 hectares = ____ acres
7. Change 12 hectares to acres.   12 hectares = ____ acres
8. How many hectares is 5 acres?   5 acres is ____ hectares
9. How many hectares is 7.5 acres? 7.5 acres is ____ hectares
10. How many hectares is 10 acres? 10 acres is ____ hectares

Ch 12 Ls 4 Measure Triangles, Trapezoids, and Rhombi 115

_____ #1 #2 ____ /10 #3 ____ / 6 R ___ / 7 T ____ / 23 _____
         Name   Checker

#1 1. How do you find the perimeter of any shape? _____
2. Given the area of a rectangle. How does a triangle change it? _____
3. What formula finds the area of any triangle? _____

4. How do you find this area?   4 meters ◿ 6 m

Multiply ___ x ___ = ___   ___ Divide by ___ = ___ sq yds

5. How do you find this area?   3 cm ◿ 9 cm

Multiply ___ x ___ = ___   ___ Divide by ___ = ___ sq yds

#2 Area of triangle, rhombus, and trapezoid

1. How do you find the area for a trapezoid? _____
2. How do you find the area of a rhombus? _____
3. How is rhombus like area of a triangle? _____

4. What is the area?   4 dm ▱ 7 dm

Multiply ___ x ___ = ___ sq yds

5. What is the area?   8 m / 4 m 4 m \ 12 m

Add ___ + ___ + ___ = ___ sq ft

#3 Find the area.   Calculator? yes no

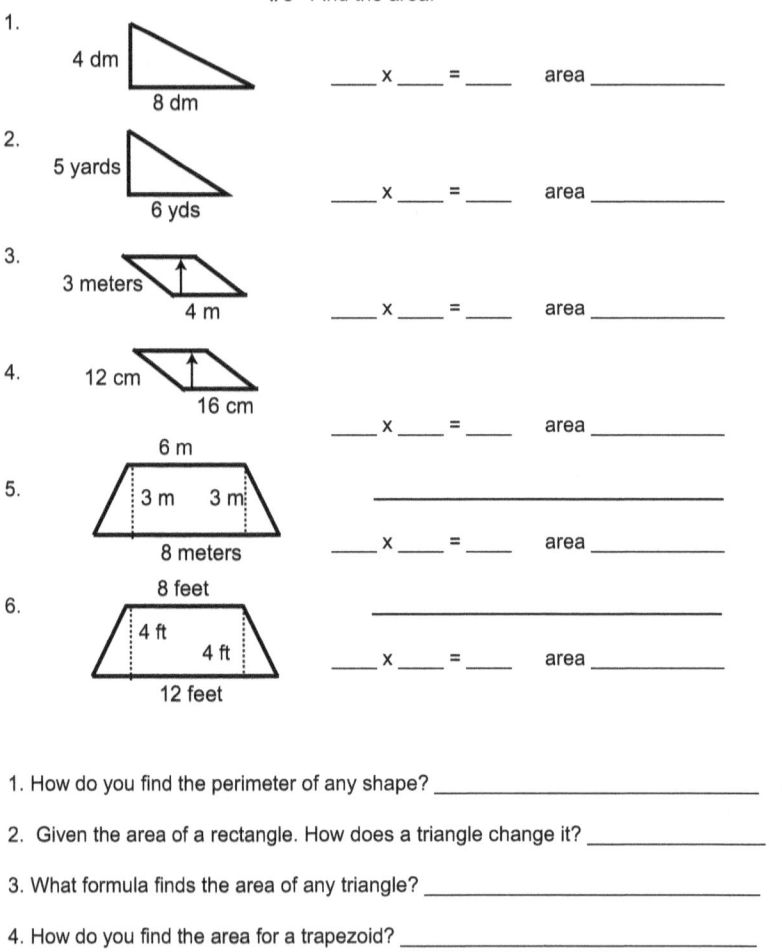

1. 4 dm, 8 dm   ___ x ___ = ___ area ___

2. 5 yards, 6 yds   ___ x ___ = ___ area ___

3. 3 meters, 4 m   ___ x ___ = ___ area ___

4. 12 cm, 16 cm   ___ x ___ = ___ area ___

5. 6 m, 3 m, 3 m, 8 meters   _____
   ___ x ___ = ___ area ___

6. 8 feet, 4 ft, 4 ft, 12 feet   _____
   ___ x ___ = ___ area ___

Review 1. How do you find the perimeter of any shape? _____   Calculator? yes no

2. Given the area of a rectangle. How does a triangle change it? _____

3. What formula finds the area of any triangle? _____

4. How do you find the area for a trapezoid? _____

5. How do you find the area of a rhombus? _____

6. Find the area of these 2 triangles.   6 yards, 17 yards   18 cm, 32 cm

_____   _____

Review Problems 117

_____ #1 #2 #3 ____/ 22 #4 #5 #6 ____/ 15 Total ____/ 37
            Name

#1 **1. Perimeter** _____
     2. Area _____
     3. Square Inch _____
     4. Square Foot _____
     5. Square Yard _____
     6. Acre _____
     7. Square Meter _____
     8. Square Centimeter _____
     9. Hectaire _____
     10. Area Triangle _____

#2 Find a perimeter and area.   Calculator?
                                                                                                                        yes   no

1. [8 meters | 4 m]   Perimeter _____
                                               Area: _____

2. [9 yards | 2 yards]   Perimeter _____
                                               Area: _____

3. [9 cm | 5 cm]   Perimeter _____
                                             Area: _____

4. [11 decimeters | 4 dm]   Perimeter _____
                                             Area: _____

#3 How do you change to square inches?   Calculator?
                                                                                                                              yes   no

1. [5 meters | 4 m]   ____ square feet
                                          Multiply ____ x ____ = ____ sq inches

2. [8 feet | 5 feet]   ____ square feet
                                          Multiply ____ x ____ = ____ sq inches

#4 How do you change to square centimeters or millimeters?   Calculator? yes no

1. [rectangle 8 m × 4 m]   _____ square meters
   Multiply _____ x _____ = _____ sq cm

2. [rectangle 7 m × 3 m]   _____ square meters
   Multiply _____ x _____ = _____ sq cm

3. [rectangle 10 m × 6 m]   _____ square meters
   Multiply _____ x _____ = _____ sq cm

4. [rectangle 15 cm × 7 cm]   _____ square cm
   Multiply _____ x _____ = _____ sq mm

5. [rectangle 7 cm × 3 cm]   _____ square cm
   Multiply _____ x _____ = _____ sq mm

#5 Find the area of these 3 triangles.   Calculator? yes no

1. triangle 5 yards × 16 yards   _____

2. triangle 10 cm × 12 cm   _____

#6 Story Problems

1. A model of a rocket has a scale of 1 cm to 3 m. If the model rocket is 30 cm tall, how tall was the rocket?   _____   Calculator? yes no

2. Mr K wants to put trees in a field. It's thirty six meters long and eight meters wide. What is the perimeter of the field in meters?   _____

3. A model of a 6-cylinder gasoline engine is built on a scale of 1 cm is 12 cm. If the length of the model engine is 8 cm, how long is the engine?   _____

Ch 13 Ls 1 Solids and Volume   119

_____ #1 #2 ____/ 12 #3 ____/ 4 R ___/ 11 T ____/ 27 _____
            Name   Checker

#1 1. What is the side of a box called? _____

2. What is the place between 2 faces called? _____

3. What is the place called between 3 edges? _____

4. What multiplies to get the volume of a box? _____

5. How is volume labeled? _____

6. How are area and volume alike? _____

#2 1. How many faces, edges, and vertices are in this box?   3 ft / 4 feet / 2 ft

2. What is the volume?   ___ faces   ___ edges   ___ vertices

___ x ___ x ___ = ___ label _____

3. What has to change to find volume?   5 meters / 4 meters / 350 cm

4. Find the volume.   Change _____ to _____.

___ x ___ x ___ = ___ label _____

5. What is the volume?   6 inches / 5 inches / 3 inches

_____

6. What is the volume?   4 cm / 6 cm / 2 cm

_____

#3 What is the volume?   Calculator?
                                                                  yes   no

1.   [box] 5 cm, 4 cm, 8 cm   ___ x ___ x ___ = _____
                                     label _____

2.   [box] 40 dm, 60 dm, 55 dm   ___ x ___ x ___ = _____
                                     label _____

3.   [box] 3 feet, 2 feet, 48 inches ___ x ___ x ___ = _____
                                     label _____

4.   [box] 7 cm, 10 cm, 15 cm   ___ x ___ x ___ = _____
                                     label _____

Review 1. What is the side of a box called? _____   Calculator?
                                                                                     yes   no
2. What is the place between 2 faces called? _____

3. What is the place called between 3 edges? _____

4. What multiplies to get the volume of a box? _____

5. How is volume labeled? _____

6. How are area and volume alike? _____

7. Find a box that has volume of 36 cubic m.   _____ x _____ x _____ = 36 cu m

8. Find a box that has volume of 40 cubic cm.   _____ x _____ x _____ = 40 cu cm

9. Find a box that has volume of 24 cubic dm.   _____ x _____ x _____ = 24 cu dm

10. Find a box that has volume of 50 cubic ft. _____ x _____ x _____ = 50 cu ft

11. Find a box that has volume of 80 cubic mm. _____ x _____ x _____ = 80 cu mm

Ch 13 Ls 2 Surface Area of a Box   121

_____ #1 #2 ____/6   #3 ____/4   R ___/6   T ____/16 _____
           Name   Checker

#1 1. What does surface area measure? _____
       2. How do you find surface area? _____
       3. How do you find the surface area of a cube? _____

#2 1. What is the surface area
          of each side?   4 meters
                                          6 meters   2 meters

                Front: ____ x ____ = ____ x 2 = ____
                Side: ____ x ____ = ____ x 2 = ____   What's the entire
                Top: ____ x ____ = ____ x 2 = ____   surface area?

                                        ____ label_____

       2. What is the surface area
          of each side?   5 cm
                                          7 cm   3 cm

                Front: ____ x ____ = ____ x 2 = ____
                Side: ____ x ____ = ____ x 2 = ____   What's the entire
                Top: ____ x ____ = ____ x 2 = ____   surface area?

                                        ____ label_____

       3. What is the surface area
          of each side?   4 ft
                                          5 feet   3 ft

                Front: ____ x ____ = ____ x 2 = ____
                Side: ____ x ____ = ____ x 2 = ____   What's the entire
                Top: ____ x ____ = ____ x 2 = ____   surface area?

                                        ____ label_____

#3 What is the surface area?   Calculator? yes   no

1.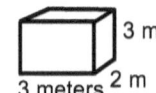
3 meters   2 m   3 m

___ + ___ + ___ = ___ x 2

_____

2.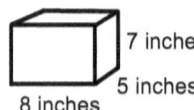
8 inches   5 inches   7 inches

___ + ___ + ___ = ___ x 2

_____

3.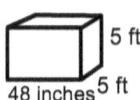
48 inches   5 ft   5 ft

___ + ___ + ___ = ___ x 2

_____

4.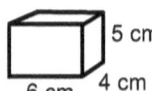
6 cm   4 cm   5 cm

___ + ___ + ___ = ___ x 2

_____

Review 1. What does surface area measure? _____   Calculator? yes   no

2. How do you find surface area? _____

3. How do you find the surface area of a cube? _____

4.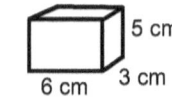
6 cm   3 cm   5 cm

___ + ___ + ___ = ___ x 2

_____

5.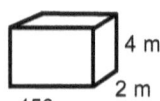
450 cm   2 m   4 m

___ + ___ + ___ = ___ x 2

_____

6.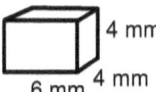
6 mm   4 mm   4 mm

___ + ___ + ___ = ___ x 2

_____

Boxes are irregular.

Ch 13 Ls 3 Triangle Prisms and Surface Area   123

_____ #1 #2 ____/ 8   #3 ____/ 8   R ___/ 10   T ____/ 26 _____
        Name   Checker

#1 1. How many faces/vertices does a triangle prism have? ____ faces ____ vertices

2. What 3 things add to get surface area of a tent? _____

3. Which 2 parts get doubled? _____

4. What's the surface area of the base and 2 sides?

4 dm ⟋⟍ 5 dm
6 dm   4 dm

What's the surface area of the end triangles?

Base ____ x ____ = ____ square dm

Side ____ x ____ = ____ x 2 = ____ square dm

What's the surface area? End 1/2 x ____ x ____ = ____ x 2 = ____ square dm

Surface area is ____ + ____ + ____ = ____ square dm

#2 Pyramid: Find Surface Area

1. How many faces does a square pyramid have? _____
2. How many vertices does a triangle pyramid have? _____
3. What 2 parts does pyramid add for surface area? _____

4. What's the area of the base and 1 side?

4 m
6 m   6 meters

What's the entire surface area?

Base ____ x ____ = ____ square m

Side 1/2 x ____ x ____ = ____ x 4 = ____ square m

Surface area is ____ + ____ = ____ square m

#3 What is the surface area?   Calculator? yes no

1. (triangular prism: 6 cm, 6 cm, 8 cm, 8.5 cm)

Base 6 x 8 = _____ sq cm
Side 8 x 6.7 = _____ x 2 = _____ sq cm
End 1/2 x 6 x 6 = _____ x 2 = _____ sq cm
Surface area is _____ + _____ + _____ = _____ sq cm

2. (triangular prism: 4 dm, 5 dm, 6 dm, 6.4 dm)

Base 5 x 6 = _____ sq dm
Side 4.7 x 6 = _____ x 2 = _____ sq dm
End 1/2 x 5 x 4 = _____ x 2 = _____ sq dm
Surface area is _____ + _____ + _____ = _____ sq dm

3. (square pyramid: 5 km, 8 km, 8 km)

Base 8 x 8 = _____ sq km
Side 1/2 x 5 x 8 = _____ x 4 = _____ sq km
Surface area is _____ + _____ = _____ sq km

4. (square pyramid: 6 m, 10 m, 10 m)

Base 10 x 10 = _____ sq m
Side 1/2 x 6 x 10 = _____ x 4 = _____ sq m
Surface area is _____ + _____ = _____ sq m

Review

1. What 3 things add to get surface area of a tent? _____   Calculator? yes no

2. Which 2 parts get doubled? _____

3. How many faces does a square pyramid have? _____

4. What 2 parts does pyramid add for surface area? _____

5. How many of these does a tent have?   ___ faces ___ edges ___ vertices

6. How many does a pyramid have?   ___ faces ___ edges ___ vertices

Ch 13 Ls 4 Triangle Prisms and Surface Area   125

_____ #1 #2 ___ / 9   #3 ___ / 8   R ___ / 9   T ___ / 26 _____
         Name   Checker

#1 1. What 3 things multiply to find volume of a triangle prism? _____
     2. How does volume of a triangle prism change volume of a box? _____
     _____

     3. What is the volume?   3 dm
                                  3 dm   4 dm

     1/2 x ___ x ___ x ___ = ___ label _____

                                  4 cm
     4. What is the volume?
                                  5 cm   6 cm

     1/2 x ___ x ___ x ___ = ___ label _____

#2 1. How many faces does a square pyramid have? _____
     2. What's the formula to find the volume of a pyramid? _____
     3. How do you compare a pyramid with the volume of it's box? _____
     _____

                                  4 m
     4. How do you find the volume?
                                  5 meters   5 meters

     $\frac{1}{3}$ x ___ x ___ x ___ = ___ label _____

                                  2 km
     5. What is the volume?
                                  3 km   3 km

     $\frac{1}{3}$ x ___ x ___ x ___ = ___ label _____

#3 What is the volume? Calculator? yes no

1. 5 m, 6 m, 8 meters

1/2 x 8 x 6 x 5 = _____
_____ label _____

2. 6 dm, 8 dm, 6 dm

1/2 x 8 x 6 x 6 = _____
_____ label _____

3. 40 cm, 50 cm, 50 cm

1/3 x 50 x 50 x 40 = _____
_____ label _____

4. 5 km, 6 km, 6 km

1/3 x 6 x 6 x 5 = _____
_____ label _____

Review 1. What 2 things multiply to find volume of a tent? _____ Calculator? yes no

2. How does volume of a triangle prism change volume of a box? _____

_____

3. How many faces does a square pyramid have? _____
4. What's the formula to find the volume of a pyramid? _____
5. How do you compare a pyramid with the volume of it's box? _____

_____

Start with the box, then find the volume.

6. Make a tent with volume 60 cubic meters   1/2 x ___ x ___ x ___ = 60 cu m

7. Make a tent with volume 50 cubic dm.   1/2 x ___ x ___ x ___ = 50 cu dm

8. Make a pyramid with volume 100 cu km   1/3 x ___ x ___ x ___ = 100 cu km

9. Make a pyramid with volume 80 cubic cm.   1/3 x ___ x ___ x ___ = 80 cu cm

Real Problems   127

_____ #1 #2 ____/11   #3 #4 ____/12 R ___/ 9 Total ____/30
        Name

#1 1. Volume _____
    2. Cube Box _____
    3. Surface Area _____
    4. Pyramid _____
    5. Triangle Prism _____

#2   Find the measurement of each figure.   Calculator?
                                                          yes   no
1.   ___ + ___ + ___ = ___ x 2
   Surface   14 cm
   area   20 cm 10 cm   _____

2.   ___ + ___ + ___ = ___ x 2
   Surface   9 ft
   area   8 ft 5 ft   _____

3.   Base 8 x 11 = ____ sq dm
   6 dm   6.8 dm
                                Side 6.8 x 11 = ____ x 2 = ____ sq dm
   Surface 8 dm 11 dm
   area   End 1/2 x 6 x 8 = ___ x 2 = ____ sq dm
                                Surface area is ____ + ____ + ____ = ____ sq dm

4.   Base 7 x 7 = ____ sq m
      6 m
   Surface   7 m   7 m   Side 1/2 x 6 x 7 = ____ x 4 = ____ sq m
   area
                                Surface area is ____ + ____ = ____ sq m

5.
   50 cm   ____ x 60 x 80 x 50 = _____
   Volume   60 cm   80 cm   ____ label _____

6.
      8 km   ____ x 12 x 12 x 8 = _____
   Volume   12 km   12 km   ____ label _____

#3 What is the surface area? (Round to 1s.)   Calculator? yes no

1. [box: 60 cm × 4 dm × 3 dm]

___ + ___ + ___ = ___ x 2

_____

2. [triangular prism: 6 m, 7.2 m, 8 m, 7 m]

Base 8 x 7 = ____ sq m
Side 7 x 7.2 = ____ x 2 = ____ sq m
End 1/2 x 7 x 7.2 = ____ x 2 = ____ sq m
Surface area is ____ + ____ + ____ = ____ sq m

[pyramid: 7 km, 9 km, 9 km]

Base 9 x 9 = _____ sq km
Side 1/2 x 7 x 9 = ____ x 4 = _____ sq km
Surface area is _____ + _____ = _____ sq km

#4 What is the volume?

1. [triangular prism: 50 cm, 70 cm, 90 cm]

1/2 x 90 x 70 x 50 = _____
____ label _____

2. [pyramid: 5 dm, 8 dm, 8 dm]

1/3 x 8 x 8 x 5 = _____
____ label _____

#5 Story Problems

1. Amav is wrapping a block of cheese that is 18 cm long by 8 cm high by 10 cm wide with wrap. What is the surface area of the cheese block?

Calculator? yes no

2. Mr C is wrapping a box that 16 dm long by 1 m wide by 5 dm high. What is the surface area of the box?

3. Keya is painting a door that is 80 cm by 200 cm by 4 cm. How much area will she paint?

4. Ojas built a model pyramid that is 20 cm high and the base is 40 cm long. What's the volume?

Ch 14 Ls 1 Parts of a Circle   129

_____ #1 #2 ____/ 10 #3 #4 ____/ 5 R ___/ 10 T ____/ 25 _____
                Name   Checker

#1 1. What does radius measure? _____

2. What does diameter measure? _____

3. If you know the radius, how do you find the diameter? _____

   4. What is the diameter of this circle?   **Radius: 7.1 meters**

                                    _____

   5. What is the radius of this circle?   **Diameter: 5.2 dm**

                                    _____

            #2   Find the perimeter of a circle.

1. What does Pi find and what is it's number? _____
2. How do you find the perimeter around a circle? _____
3. What part of the circle changes the perimeter? _____

   4. What's the 1st step to
      find the perimeter?   Radius 4 cm

                    Multiply ___ x 2 = ___ cm   What is the perimeter?

                              3.1 x ___ = ___ cm

   5. All 1 step, what's the perimeter
      if the radius is 5 km?   Radius 5 km

              Multiply ___ x 2 = _____ km   3.1 x ___ = ___ km

#3 What is the diameter of this circle?   Calculator? yes   no

1. Find the diameter.

Radius: 3.6 meters

Radius: 9.8 meters

_____   _____

#4 What is the perimeter of this circle?

1. Find the perimeter with radius 20 dm.   Radius 20 dm   _____

2. Find the perimeter with radius 8 cm.   Radius 8 cm   _____

3. Find the perimeter with radius 7 km.   Radius 7 km   _____

Review 1. What does the radius measure? _____   Calculator? yes   no

2. What does diameter measure? _____

3. If you know the radius, how do you find the diameter? _____

4. What does Pi find and what is it's number? _____

5. How far is it all the way around a circle using pi? _____

6. How do you find the perimeter around a circle? _____

7. What part of the circle changes the perimeter? _____

8. What radius has a perimeter of 30 dm? _____

9. What radius has a perimeter of 40 cm? _____

10. What radius has a perimeter of 60 meters? _____

Ch 14 Ls 2 Cans and Surface Area   131

_____ #1 #2 ____ / 10 #3 ____ / 2 R ____ / 9 T ____ / 21 _____
              Name   Checker

#1 1. What formula finds the area of a circle? _____
       2. If the radius is 2 meters, find the area for the circle. **3.1 x ___ = ___ square meters**
       3. What's the area of a square that is 4 m x 4 m? _____
       4. What is the difference between the square and circle? _____

       5. Find the area of the circle. First step?   Radius:
                                                           3 meters

       _____

       _____

#2 1. How many faces and edges does a can have? _____
       2. What is hidden information? _____
       3. What 2 things add to find the surface area of a cylinder? _____

       _____

       4. What is the area of the circle?   3 cm
                                                    2 cm

                                                              What do you use to find
                3.1 x _____ = _____   the area of the rectangle?
                       radius   area

                              Perimeter is 3.1 x _____ = _____
       5.   It's the diameter.
              What's the surface area?   3 x _____ = _____ _____
                                         height diameter area   label

                              _____ + _____ + _____ = _____ square centimeters

#3 Find the area of a circle.   Calculator?
                                                                   yes   no

1. Radius: 6 meters _____

   _____

2. Radius: 40 cm _____

   _____

Review 1. What formula finds the area of a circle? _____ Calculator?
                                                                              yes   no
2. If the radius is 2 inches, find the area for the circle. **3.1 x ___ = ___ sq inches**

3. What's the area of a square that is 4 in x 4 in? _____

4. What is the difference between the square and circle? _____

5. How many faces and edges does a can have? _____

6. What is hidden information? _____

7. What 2 things add to find the surface area of a cylinder? _____

_____

8. Find the surface area of the can.
   What is the area of the base and top?   4 meters
                         3 meters

   3.1 x _____ = _____
        radius squared   area

   Find the area of the rectangle. Perimeter is 3.1 x _____ = _____

   Find the surface area of the can.   _____ + _____ + _____ = _____ square meters

9. Find the surface area of the can.
   What is the area of the base and top?   8 inches
                         2 inches

   3.1 x _____ = _____
        radius squared   area

   Find the area of the rectangle. Perimeter is 3.1 x _____ = _____

   Find the surface area of the can.   _____ + _____ + _____ = _____ square inches

Ch 14 Ls 3 Surface Area of a Cone   133

_____   #1 #2 ____ / 7   #3 ____ / 4 R ____ / 4 T ____ / 15 _____
　　　　Name　　　　　　　　　　　　　　　　　　　　　　　　　　　　　　　　Checker

#1 1. How many faces, edges, and vertices are in a cone? _____

2. Name 3 steps to find surface area of a cone when you know the side of it. _____

_____

3. What formula finds the outside if you know the height of the cone? _____

_____

#2 1. What is the area of the base circle?   4 meters

　　　　　　　　　　　　　　　　　3 meters

　　　　3 m　　　3.1 x _____² = _____ _____　　What do you use to find
　　　　　　　　　　　radius　　area　label　　the area of the triangle?
　　　　　　　　　　　squared

2.　　　　　　　Per 3.1 x 6 = _____ _____
　　　　　　　　　　　　　　　　　　　　　　　　　What is the
　4 m　　　1/2 x _____ x 4 = _____ _____　　surface area?
　　　m　　length width　　　　label

　　　　　　　_____ + _____ = _____ _____
　　　　　　　base　triangle　　　label

3. What is the area of the base circle?   6 inches

　　　　　　　　　　　　　　　5 in

　　5 in　　3.1 x _____² = _____ _____　　What do you use to find
　　　　　　　radius　area　label　　the area of the triangle?
　　　　　　　squared

4.　　　　　　Per 3.1 x 10 = _____ _____
　　　　　　　　　　　　　　　　　　　　　　　　What is the
　6 in　　　0.5 x _____ x _____ = _____ _____　surface area?
　　　in　　length　width　　　label

　　　　　_____ + _____ + _____ = _____ _____
　　　　　_base　top　rectangle　　　label

#3 Find the surface area of a cone.   Calculator? yes no

1. Find the area of the base circle.

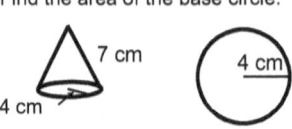

3.1 x ___^{2} = ___ ___
    radius   area label
    squared

Per 3.1 x 4 = ___ ___

What do you use to find the area of the triangle?

What is the surface area?

0.5 x ___ x ___ = ___ ___
      length width   label

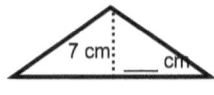

___ + ___ + ___ = ___ ___
base top rectangle   label

2. What is the area of the base circle?

3.1 x ___^{2} = ___ ___
    radius   area label
    squared

Per 3.1 x 8 = ___ ___

What do you use to find the area of the triangle?

What is the surface area?

1/2 x ___ x ___ = ___ ___
      length width   label

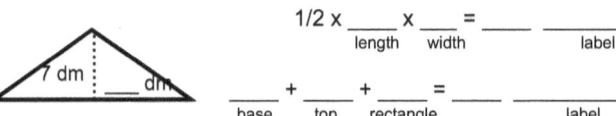

___ + ___ + ___ = ___ ___
base top rectangle   label

Review 1. How many faces, edges, and vertices are in a cone? ___   Calculator? yes no

2. Name 3 steps to find surface area of a cone when you know the side of it. ___

___

3. What formula finds the outside if you know the height of the cone? ___

___

4. If a cone has a height of 2 m and base of 60 cm, which surface area do you choose?

___

Ch 14 Ls 4 Finding Can/Cone Volume 135

_____ #1 #2 ____/ 9   #3 ____/ 8   R ___/ 6   T ____/ 23 _____
             Name   Checker

#1 1. What's the 1st step to find the volume of a cylinder? _____
    2. What does the base multiply with? _____
    3. What's the formula for volume of a cylinder? _____
        4. What is the area of
           the base and top?   4 m

                             3 m

        Pi R squared   3.1 x ____^{2} = ____ square meters   What is the
                                                                       volume of the can?

                          ____ x ____ = ____ cubic meters

                       #2 Find the Volume of a Cone

    1. What's the 1st step to find the volume of a cone? _____
    2. What does the base multiply with? _____
    3. How is the volume of a cone different from a can? _____
    4. What is the formula for volume of a cone? _____

        5. What is the area of the base?   Height 5 meters
           Find the 1st step.
                                            3 meters

                                                              What does if
             3 m   multiply for volume?
                    3.1 x _____ = _____   _____
                         radius   area   label

    Multiply 1 third $\frac{1}{3}$ x ____ x ____ = ____   _____
                             base   height   label

#3 Name 2 things find the volume of a can.   Calculator? yes   no

1. **Radius: 2 centimeters**   3.1 x ____^{2} = ____ x ____ = ____ sq cm
   Height: 3 centimeters   ____ x ____ = ____ cubic cm
   What's the cone volume?   ____ x ____ = ____ cubic cm

2. **Radius: 1 meters**   3.1 x ____^{2} = ____ x ____ = ____ sq m
   Height: 4 meters   ____ x ____ = ____ cubic m
   What's the cone volume?   ____ x ____ = ____ cubic m

3. **Radius: 3 feet**   3.1 x ____^{2} = ____ x ____ = ____ sq ft
   Height: 5 feet   ____ x ____ = ____ cubic ft
   What's the cone volume?   ____ x ____ = ____ cubic ft

4. **Radius: 4 dm**   3.1 x ____^{2} = ____ x ____ = ____ sq dm
   Height: 10 dm   ____ x ____ = ____ cubic dm
   What's the cone volume?   ____ x ____ = ____ cubic dm

Review 1. What's the 1st step to find the volume of a cylinder? _____   Calculator?
2. What the circle area multiply with to find volume? _____   yes   no
3. What's the formula for volume of a cylinder? _____
4. What's the 1st step to find the volume of a cone? _____
5. What fraction does it multiply with to find the volume of a cone? _____
6. What is the formula for volume of a cone? _____

Ch 14 Ls 5 Volume/Surface Area of a Sphere   137

_____ #1 #2 ____ / 7 #3 ____ / 3 R ___ / 8 T ___ / 18 _____
           Name   Checker

#1 1. How many faces, edges, and vertices are in a sphere? _____
    2. What's the formula for surface area of a sphere? _____
    3. What's the equation for a sphere if you use a number for pi? _____

    _____

    4. Find the volume of a 2 meter sphere.
       What is the formula? (Use the number)   2 m

    What's the answer? $4.2 (\quad)^3 = 4.2 \times$ _____

    _____ _____
              label

#2 Surface Area of a Sphere

    1. What's the formula for the surface area of a sphere? _____
    2. Name 2 ways it changes the volume of a sphere. _____
    3. What's the number formula for the surface area? _____

    4. Find the surface area of a 2 meter sphere.
       What is the formula? (Use the number)   2 m

         $12.4 (\quad)^2 = 12.4 \times$ _____   What's the answer?

    _____ _____
              label

#3 Find the surface area of these spheres.   Calculator? yes no

1.  5 m $12.6\ (\quad)^2 = 12.6 \times$ ____

   _____ _____
              label

2. 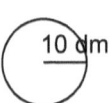 10 dm $12.6\ (\quad)^2 = 12.6 \times$ ____

   _____ _____
              label

3. 20 cm $12.6\ (\quad)^2 = 12.6 \times$ ____

   _____ _____
              label

Review   1. How many faces, edges, and vertices are in a sphere? _____   Calculator? yes no

2. What's the formula for surface area of a sphere? _____

3. What's the equation for a sphere if you use a number for pi? _____

_____

4. What's the formula for the surface area of a sphere? _____

5. Name 2 ways it changes the volume of a sphere. _____

6. What's the number formula for the surface area? _____

Find the volume of these spheres.

7. 3 m $4.2\ (\quad)^3 = 4.2 \times$ ____

   _____ _____
              label

8. 4 cm $4.2\ (\quad)^3 = 4.2 \times$ ____

   _____ _____
              label

Review Problems 139

_____ #1 #2 ____/11 #3 #4 ____/ 6 R ___/ 7 T ____/ 23
          Name

#1 1. Circle _____
    2. Radius _____
    3. Diameter _____
    4. Pi _____
    5. Can _____
    6. Cone _____
    7. Sphere _____

#2 Find these circle measurements.   Calculator?
                                                              yes   no

1. Find the perimeter
   and area radius 5 ft.   Radius 5 feet   _____

2. Find the perimeter
   and area radius 9 ft.   Radius 9 feet   _____

3. Find the surface area of the can.   6 inches
   What is the area of the base and top?
                                       2 inches

   3.1 x ____ = ____
         radius squared   area

   Find the area of the rectangle. Perimeter is 3.1 x ____ = ____
   Find the surface area of the can.   ____ + ____ + ____ = ____ sq inches

4. Find the area of the base circle.

                                          3.1 x ____² = ____ ____
Find the surface   7 feet   3 ft   radius   area   label
area of a cone.   squared
         3 feet
                                          Per 3.1 x 6 = ____ ____

   What do you use to find   What is the surface area?
   the area of the triangle?
                                   0.5 x ____ x ____ = ____ ____
                                         length width   label
           7 ft
              ____ ft   ____ + ____ + ____ = ____ ____
                                   base   top rectangle   label

#3 Find the volume.   Calculator? yes no

1. Radius: 4 feet
Height: 8 feet
Volume

$3.1 \times \underline{}^2 = \underline{} \times \underline{} = \underline{}$ sq yd

$\underline{} \times \underline{} = \underline{}$ cubic yd

What's the cone volume? $\underline{} \times \underline{} = \underline{}$ cubic yd

2. (circle, 6 ft)
Volume
Surface area

$4.2 ()^3 = 4.2 \times \underline{}$ $\underline{}\ \underline{}$ label

$12.4 ()^2 = 12.4 \times \underline{}$ $\underline{}\ \underline{}$ label

3. (circle, 7 cm)

$4.2 ()^3 = 4.2 \times \underline{}$ $\underline{}\ \underline{}$ label

$12.4 ()^2 = 12.4 \times \underline{}$ $\underline{}\ \underline{}$ label

#4 Story Problems   Calculator? yes no

1. A monster truck uses 23 degree tires 65 inches tall, mounted on 25-inch diameter wheels. What is the circumference of a monster truck wheel?

2. On June 8, 1992, a crop circle with an 18-meter radius was found in a wheat field. Estimate its circumference.

3. Find out how far the needle travels in one rotation of a vinyl disc that is 30 centimeters across. (One turn of the outside of the disc.)

4. A pyramid is 85 meters high, each side is 11 m, and has an estimated weight of 1,8 million kilograms. Find the pyramid's circumference.

5. A can of tomatoes is 15 cm tall and has a radius of 5 cm. What is it's surface area?

6. A cone has height of 60 cm and has a base that's 10 cm radius. What is it's volume?

7. A beach ball has a radius of 25 cm. What's the volume?

www.ingramcontent.com/pod-product-compliance
Lightning Source LLC
Chambersburg PA
CBHW030753180526
45163CB00003B/1000